# TOPOLOGICA
# AN
# DIFFERENTIA



GEORGE R. SELL
*University of Minnesota*
*Minneapolis, U.S.A.*



VAN NOSTRAND REINHOLD COMPANY
LONDON

NEW YORK · CINCINNATI · TORONTO · MELBOURNE

VAN NOSTRAND REINHOLD COMPANY
Windsor House, 46 Victoria Street, London S.W.1

INTERNATIONAL OFFICES
New York  Cincinnati  Toronto  Melbourne



Library of Congress Catalog Card No. 74–141983
ISBN 0 442 07502 2

First published 1971

Printed in Great Britain by
Butler & Tanner Ltd
Frome and London

Dedicated to *my Parents*

# PREFACE

These notes are based on lectures given at the University of Minnesota and the University of Southern California during 1966 and 1967. They represent the author's view of the theory of topological dynamics with special emphasis on applications to nonautonomous differential equations.

The main construction, which allows one to apply the theory of topological dynamics (in a meaningful way) to nonautonomous differential equations is given in Chapter IV. Applications of this construction are presented in Chapters VIII, IX and X. These applications are presented, not as a complete theory, but rather as an illustration of how the theory of topological dynamics can be exploited in the study of differential equations.

These notes are written for the person who has some background in the basic existence and uniqueness theorems of ordinary differential equations and some knowledge of elementary point set topology. For the later chapters (Chapters IX, X, XI) some knowledge of integration theory and functional analysis will also be needed.

Essentially, the notes are self-contained. References are given for results not proved here.

The author would like to thank all his colleagues and students who by their comments and suggestions helped me in my own understanding of this subject. I am particularly grateful to Professors H. A. Antosiewicz, J. Auslander, W. A. Harris, Jr., W. T. Kyner, R. K. Miller, R. Sacker, G. Seifert and Y. Sibuya.

## CONTENTS

# CHAPTER I

## *REVIEW OF UNIFORM SPACES*

In this chapter we shall review some of the basic concepts from the theory of uniform spaces. We shall assume that the reader has a background one would normally achieve after studying the topology of metric spaces. The purpose of this chapter is then to introduce the terminology and notation of "generalized sequences" and "uniform topologies." The reader is referred to J. Kelley [16] for a more detailed discussion and for the proofs of the statements we make below.

I.A. **Generalized sequences.** A **directed set** $(A, \geq)$ is defined to be a nonempty set $A$ with an ordering $\geq$ that satisfies the following conditions:

(i) (Transitivity) If $a, b$ and $c$ are in $A$ and $a \geq b$ and $b \geq c$, then $a \geq c$.

(ii) (Reflexivity) If $a$ is in $A$, then $a \geq a$.

(iii) (Upper bound) For every pair $b$ and $c$ in $A$ there is an $a$ in $A$ such that $a \geq b$ and $a \geq c$.

We sometimes denote the directed set by the symbol $A$.

A **(generalized) sequence** is a function $f$ defined on a directed set $A$ with range in another set. (A generalized sequence is referred to as a "net" in [16].) The value of a generalized sequence $f$

at a point $a$ in $A$ is denoted by $f_a$. A generalized sequence $f$ is said to be **in** $X$ if the range of $f$ is in $X$, that is $f_a \in X$ whenever $a \in A$. A generalized sequence is said to be **eventually in** $U$, where $U \subset X$, if there is a $b \in A$ such that $f_a \in U$ whenever $a \geq b$.

Let $f: A \to X$ be a generalized sequence, where $X$ is a topological space. The sequence $f$ is said to **converge** to a point $x$ in $X$ if it (the sequence) is eventually in every neighborhood of $x$. (This concept of convergence is the same as the "Moore-Smith convergence" which is treated at length in [16].) We sometimes write this as $f_a \to x$, or $\lim_a f_a = x$.

If $A$ is the set of natural numbers $\{1, 2, ...\}$ with the usual ordering, then a generalized sequence is an *ordinary sequence,* and the concept of convergence is equivalent to the concept of "sequential convergence."

The importance of this concept of convergence of generalized sequences, for our purposes, is that it gives a convenient way of describing the topology on a set $X$. (We refer the reader to [16, Chap. II] for a more detailed discussion of this relationship. The same relationship can be constructed by using the notion of a "filter" cf. [5].) The following statement summarizes the relationship between convergence (of generalized sequences) and topologies.

I.1. PROPOSITION. *Let $X$ be a topological space and $U \subset X$. A point $x$ is in* Cl$U$ *if and only if there is a generalized sequence $\{x_a\}$ in $U$ that converges to $x$.*

If a topological space $X$ satisfies the first axiom of countability, that is, if every point $x$ in $X$ has a countable neighborhood basis, then the last proposition can be replaced by an ordinary sequence. We shall be particularly interested in this when the space $X$ is metrizable.

I.2. PROPOSITION. *Let $X$ be a metrizable topological space and $U \subset X$. Then a point $x$ is in* Cl $U$ *if and only if there is an (ordinary) sequence $\{x_n\}$ in $U$ that converges to $x$.*

In the sequel we shall generally drop the modifiers "generalized" or "ordinary" when discussing sequences. It should be clear from the context which concept is required.

There is a special type of convergence which will be used quite often in the sequel. Let $t: A \to R$ (or $t: A \to I$) be a sequence, where $R$ denotes the reals and $I$ the integers. We shall say that $t_a \to +\infty$, or $t_a \to -\infty$, if the sequence $\{t_a\}$ converges to $+\infty$, or $-\infty$, in the two-point compactification of $R$ (or $I$) given by

$$R \cup \{+\infty\} \cup \{-\infty\} \quad (\text{or } I \cup \{+\infty\} \cup \{-\infty\}).$$

In other words, $t_a \to +\infty$ if for every integer $N$ there is an index $b$ in $A$ such that $t_a \geq N$ whenever $a \geq b$.

We shall need the concept of a (generalized) subsequence. Let $A$ be a directed set and $B \subset A$. We shall say that $B$ is a **cofinal subset** of $A$ if $B$ is a directed set in the same ordering, and for each $a \in A$ there is a $b \in B$ with $b \geq a$. If $f: A \to X$ is a sequence and $B$ is a directed subset of $A$, then the restriction of $f$ to $B$, $f: B \to X$, is said to be a (**generalized**) **sequence.**

I.B. **Iterated Limits.** Let $A_1$ and $A_2$ be two directed sets and let $A = A_1 \times A_2$ be the Cartesian product. The set $A$ becomes a directed set when we define $(a_1, a_2) \geq (b_1, b_2)$ to hold whenever $a_1 \geq b_1$ and $a_2 \geq b_2$. With this ordering $A$ is referred to as the **directed product set.** In general, if one has a family $(A_x, \geq_x)$ where $x$ lies in some index set $X$ we can define an ordering on the cartesian product $A = \times_x A_x$ by saying that $a \geq b$ provided that for each component one has $a_x \geq_x b_x$ for $x \in X$.

We now state the theorem on iterated limits.

I.3. PROPOSITION. *Let* $D$ *be a directed set. For each* $n \in D$, *let* $A_n$ *be a directed set. Let* $A = \times_n A_n$ *be the directed product sets generated by* $\{A_n : n \in D\}$ *and let* $B = D \times A$ *be the directed product set generated by* $D$ *and* $A$. *Let* $R$ *be defined on* $B$ *by*

$$R : (n, a) \to (n, a_n) \in D \times A_n .$$

*For each* $n$ *in* $D$, *let* $S(n, \cdot) = S(n, a_n)$ *be a sequence defined on* $A_n$ *with range in a fixed topological space* $X$. *Then* $S \circ R$ *is a sequence defined on* $B$ *with range in* $X$ *and* $S \circ R$ *converges to* $\lim_n \lim_{a_n} S(n, a_n)$ *whenever this iterated limit exists.*

I.C. **Uniform Topologies.** A **uniform neighborhood system** for a set set $X$ is defined to be a quadruple $(X; A, \geq; V)$ where $X$ is a set, $(A, \geq)$ a directed set and for each $a \in A$ and $x \in X$, $V_a(x)$ is a subset of $X$ subject to the following conditions:

(i) $x \in V_a(x)$ ;

(ii) $V_a(x) \subset V_b(x)$ whenever $a \geq b$ ;

(iii) (Symmetry condition.) $y \in V_a(x)$ if and only if $x \in V_a(y)$ ;

(iv) (Uniformity condition.) For each $a$ in $A$ there is a $b$ in $A$ such that $z \in V_a(x)$ whenever $z \in V_b(y)$ and $y \in V_b(x)$.

Condition (iv) is referred to as a uniformity condition since it asserts that the choice of $b$ depends only on $a$. By successive applications of the uniformity condition one can easily establish the following result.

I.4. PROPOSITION. *Let* $a \in A$ *and let* $n \geq 1$ *be an integer. Then there is a* $b$ *in* $A$ *such that*

$$x_i \in V_b(x_{i+1}),\ i = 1, \ldots, n \Longrightarrow x_1 \in V_a(x_{n+1}) .$$

A uniform neighborhood system for a set $X$ generates a topology in the usual fashion. That is, a set $U \subset X$ is said to be **open** if for every $x$ in $U$ there is a neighborhood $V_a(x) \subset U$. A set $U \subset X$ is **closed** if $X - U$ is open. This topology is called the **uniform topology** (generated by $A$ and $V$.) The set $X$ with their topology is referred to as a **uniform space.** We shall denote this in any of the following ways: $X$, $(X, A)$, $(X, A, V)$ or $(X, A, \geq, V)$.

There is no loss in generality in assuming that the sets $V_a(x)$ are open in the uniform topology. Indeed, if we define $V_a^0(x)$ to be the largest open set in $V_a(x)$ then one can show that $(X, A, \geq, V^0)$ is a uniform neighborhood system for $X$ and the topology generated by $A$ and $V^0$ agrees with that generated by $A$ and $V$. We leave the proof of these assertions as an exercise. Furthermore, in the sequel we will assume that the sets $V_a(x)$ are open. The reader should observe that in the four examples in Section I.E, the constructed sets $V_a(x)$ are indeed open.

One should note that a topological space $X$ is a Hausdorff space (that is, distinct points have disjoint neighborhoods) if and only if every sequence in $X$ has at most one limit point. For a uniform space $(X, A, V)$ this is equivalent to asserting that for any two distinct points $x$ and $y$ in $X$, there is an index $a$ in $A$ such that $V_a(x) \cap V_a(y) = \emptyset$. We shall assume in the sequel that the space $(X, A, V)$ is a Hausdorff space.

For our purposes, then, a uniform neighborhood system in addition to satisfying the four conditions above, will satisfy

(v) each $V_a(x)$ is open;

(vi) if $x$ and $y$ are distinct points in $X$, then there is an $a$ in $A$ such that $V_a(x) \cap V_a(y) = \emptyset$.

I.5. PROPOSITION. *Let $(X, A, V)$ be a uniform space. Then for every $a$ in $A$ there is a $b$ in $A$ such that* $\mathrm{Cl}\, V_b(x) \subset V_a(x)$ *for all* $x$ *in* $X$.

For a proof of this see [16, p. 179ff].

By **compactness** we mean Heine-Borel compactness. A set $U$ in $X$ is **conditionally compact** if the closure $\mathrm{Cl}\, U$ is compact. We note that a set $U \subset X$ is compact if and only if every generalized sequence in $U$ has a convergent subsequence, cf. [16, p. 136].

A sequence $\{x_n\}$ in a uniform space $(X, A, V)$ is said to be a **Cauchy sequence** if for every $a$ in $A$ there is an index $N$ such that $x_n \in V_a(x_m)$ for all $n, m \geq N$. A uniform space is said to be **complete** if every Cauchy sequence in $X$ is convergent. We note that every uniform space has a completion, cf. [16, p. 196].

A set $U \subset X$ is said to be **totally bounded** in $(X, A, V)$ if for every $a$ in $A$ there is a finite collection $\{x_1, x_2, \ldots, x_n\}$ in $U$ such that

$$U \subset \cup_i V_a(x_i) \ .$$

We note that a uniform space is compact if and only if it is totally bounded and complete, cf. [16, p. 198].

Let $f: X \to Y$ be a function where $(X, A, V)$ and $(Y, A', V')$ are uniform spaces. Then $f$ is said to be **uniformly continuous** if for every index $a'$ in $A'$ there is a $b$ in $A$ such that $f(x) \in V_a'(f(z))$ whenever $x \in V_b(z)$.

A sequence $\{f_n\}$ of functions $f_n : D \to X$ of a set $D$ into a uniform space $X$ is said to **converge uniformly** to a function $f$ if for every $a$ in $A$ there is an index $N$ such that $f_n(y) \in V_a(f(y))$ for all $n \geq N$ and all $y \in D$.

A refinement of Proposition I.1 will be convenient for some applications.

I.6. PROPOSITION. *Let $(X, A, V)$ be a uniform space and let $U \subset X$. A point x is in* Cl $U$ *if and only if there is a generalized sequence $\{x_a\}$, defined on $A$ with range in $U$, such that $x_a \to x$.*

Another property we need is the following:

I.7. PROPOSITION. *Let $(X, A, V)$ be a uniform space and let $\{x_n\}$ be a sequence in $X$ with $x_n \to x$. Then for any $a \in A$ one has the following property: If $y \in V_a(x)$, then there is an index $N$ such that $y \in V_a(x_n)$ for $n \geq N$.*

This is a consequence of the fact that the neighborhoods $V_a(x)$ are open.

I.D. **Linear topological spaces**. If $X$ is a linear space, then for some applications we will be interested in a stronger uniform structure.

We shall say that the uniform neighborhood system $(X, A, \geq, V)$ gives a **uniformly linear structure** on the linear space $X$ if, in addition to the six properties listed in Section I.C, the following are satisfied:

(vii) $V_a(x) = x + V_a(0)$ for all $a \in A$, $x \in X$.

(viii) For every $a$ in $A$ there is a $b$ in $A$ such that if $x, y \in V_b(0)$, then $x + y \in V_a(0)$.

(ix) If $x \in V_a(0)$, then $-x \in V_a(0)$.

Property (vii) asserts that the neighborhoods are invariant under translation. Property (viii) implies that addition is uniformly continuous, and property (ix) implies that inverses are continuous.

I.E. **Examples** 1. *Metric spaces.* Let $X$ be a space with a metric $d$ and let $A = \{a \in R : a > 0\}$ with the usual ordering. Define $V_a(x)$, for $a \in A$ and $x \in X$, by

$$V_a(x) = \{y \in X : d(x, y) < 1/a\} .$$

The quadruple $(X, A, \geq, V)$ forms a neighborhood system for $X$ and the uniform topology is the metric topology.

2. *Weak topologies.* Let $X$ be a Banach space with $X^*$ as its conjugate space. Let $A$ be the collection of all objects of the form $a = (x_1^*, \ldots, x_n^* ; \alpha)$ where $x_i^* \in X^*$, $1 \leq i \leq n$, ($n$ depends on $a$) and $\alpha > 0$. Define an ordering on $A$ by

$$a = (x_1^*, \ldots, x_n^* ; \alpha) \geq (y_1^*, \ldots, y_m^* ; \beta) = b$$

provided $\alpha \geq \beta$ and the collection $\{x_1^*, \ldots, x_n^*\}$ contains the collection $\{y_1^*, \ldots, y_m^*\}$. Define

$$V_a(0) = \{x \in X : |x_i^*(x)| < 1/\alpha ,\ 1 \leq i \leq n\}$$

where $a = (x_1^*, \ldots, x_n^* ; \alpha)$. Let $V_a(x) = x + V_a(0)$. Then $(A, \geq, V)$ defines a uniformly linear structure on $X$ and the uniform topology is the weak topology.

One should note that if the dimension of $X$ is infinite, then the weak topology is nonmetrizable and this topology does not satisfy the first axiom of countability, cf. [9].

3. *Weak *-topologies.* Let $X$ be a Banach space that is the conjugate of another Banach space $Y$, that is $X = Y^*$. In the same fashion as presented in Example 2 one can construct a uniform topology for $X$ that agrees with the weak* -topology for $X$. A typical element of the index set $A$ is $a = (y_1, \ldots, y_n ; \alpha)$ where $y_i \in Y$, $1 \leq i \leq n$, and $\alpha > 0$. The set $V_a(0)$ is defined by

$$V_a(0) = \{x \in X : |x(y_i)| < 1/a, \quad 1 \leq i \leq n\}$$

4. *Strong topologies.* Let $X$ and $Y$ be Banach spaces and let $L(X, Y)$ be the space of bounded linear operators from $X$ into $Y$. The strong topology on $L(X, Y)$ is defined (in terms of generalized sequences) as follows: One says that $T_n \to T$ if for every $x$ in $X$ one has $T_n x \to Tx$ in $Y$. This topology can be given a uniformly linear structure as follows: Let $A$ be the collection of all objects of the form $a = (x_1, \ldots, x_n ; a)$ where $x_i \in X$, $1 \leq i \leq n$, and $a > 0$. Use the same ordering as defined in Example 2. Define $V_a(0)$ by

$$V_a(0) = \{S \in L(X, Y) : \|Sx_i\| < 1/a, \ 1 \leq i \leq n\}$$

and $V_a(T) = T + V_a(0)$.

*Exercises* 1. Define a weak topology on the space $L(X, Y)$ in Example 4 as follows: Say that $T_n \to T$ if for each $x$ in $X$ and $y^*$ in $Y^*$ one has $y^*(T_n x) \to y^*(Tx)$. Show that this topology can be given a uniformly linear structure.

## CHAPTER II

## *BASIC PROPERTIES OF DYNAMICAL SYSTEMS*

In this chapter we shall give the traditional definition of a dynamical system, or flow, and discuss some of the elementary properties.

### II.A. **Flows and Semi-flows.**

Throughout this chapter we shall assume that $X$ is a uniform space with a Hausdorff topology generated by a directed set $(A, \geq)$ and a correspondence $V$. Many of our statements will be valid in even greater generality, but this viewpoint will be adequate.

A **dynamical system** *on* $X$ is defined to be a mapping

$$\pi: X \times T \to X$$

where $T$ is a topological group subject to the conditions:

(i) (Identity property); $\pi(x, 0) - x$.

(ii) (Group property); $\pi(\pi(x, t), s) = \pi(x, t+s)$.

(iii) $\pi$ is continuous.

In these notes we will only be interested in the two groups $R$, the reals, and $I$, the integers, with the usual topologies. The dynamical system $\pi$ is sometimes referred to as a **(continuous) flow** when $T = R$, and a **(discrete) flow** when $T = I$. In the sequel we shall use the symbol $T$ to refer to either $R$ or $I$. Many of our statements will be valid in either situation.

et $\pi$ be a flow on $X$ and define $\pi_t(x) = \pi(x, t)$.

II.1 PROPOSITION. *For each $t$ in $T$, the mapping $\pi_t$ is a homeomorphism of $X$ onto $X$, that is a bijection.*

*Proof:* The group property (ii) implies that $(\pi_t)^{-1} = \pi_{-t}$, and therefore both $\pi_t$ and $(\pi_t)^{-1}$ are continuous by the continuity property (iii). Furthermore, $\pi_t$ is an injection. Indeed, if $\pi_t(x_1) = \pi_t(x_2)$, then by the group property one has

$$x_1 = \pi(\pi(x_1, t), -t) = \pi(\pi(x_2, t), -t) = x_2 .$$

Finally, $\pi_t$ is surjective for if $y \in X$, then $y = \pi_t(x)$ where $x$ is given by $x = \pi(y, -t)$. QED.

Discrete flow can be characterized in terms of homeomorphism as the following statement asserts:

II.2 PROPOSITION. *A necessary and sufficient condition that a mapping $\pi: X \times I \to X$ be a discrete flow is that there exists a homeomorphism $F: X \to X$ of $X$ onto $X$ such that*

$$\pi(x, n) = F^n(x) , \qquad n \in I ,$$

*where $F^0$ is defined to be the identity.*

Therefore the study of discrete flows is equivalent to the study of homeomorphisms and their iterates.

If $\pi: X \times R \to X$ is a continuous flow on $X$ and we define $F$: $X \to X$ by $F(x) = \pi(x, 1)$ then $F$ is a homeomorphism of $X$ onto $X$ and $\pi(x, n) = F^n(x)$ defines a discrete flow on $X$. In this case we say that the homeomorphism $F$ is **generated by** the continuous flow $\pi$. The problem of characterizing those homeomorphisms that are generated by continuous flows is basically unresolved. We shall discuss this further in the exercises.

If $T$ is $R$ or $I$, we define $T^+ = \{t \in T : t \geq 0\}$, so $T^+$ is then a semi-group. We define a *semi-flow* as a mapping $\pi : X \times T^+ \to X$ satisfying

(i) (Identity Property $\pi(x, 0) = x$;

(ii) (Group Property) $\pi(\pi(x, t), s) = \pi(x, t+s)$;

(iii) $\pi$ is continuous.

We see then that by restricting $t$ to $T^+$ any flow gives rise to a semi-flow. The converse is not true simply because one may be unable to "back-up" a semi-flow. In other words, if $\pi$ is a semi-flow and $t \in T$, the inverse mapping $(\pi_t)^{-1}$ may fail to be defined, or if defined, it may fail to be continuous. As a result of this, Proposition 1 is no longer valid. Proposition 2 takes on the following form:

II.3. PROPOSITION. *A necessary and sufficient condition that a mapping $\pi : X \times I^+ \to X$ be a discrete semi-flow is that there exist a continuous mapping $F : X \to X$ (not necessarily surjective) such that*

$$\pi(x, n) = F^n(x), \qquad n \in I^+.$$

We see then that the study of discrete semi-flows is equivalent to the study of continuous mappings and their iterates.

II.B. **An important example.** We shall consider a large class of examples in the next chapter. Here we want to examine a basic example to show the inadequacy of the above definition of a flow, and to motivate a modified definition.

Consider the differential equation

$$x' = f(x) \tag{1}$$

on Euclidean space $R^n$ where $f$ is a $C^1$-function. Then for every point $x \in R^n$, there is one and only one solution $\phi(x, t)$ of (1) that satisfies the initial value problem $\phi(x, 0) = x$. It would appear that one could define a continuous flow $\pi$ on $R^n$ by setting $\pi(x, t) = \phi(x, t)$, and indeed this is the case *provided* every solution $\phi(x, t)$ can be continued for all time $t$. As is well known, though, this global existence property is not shared by all differential equations. The scalar equation

$$x' = x^2$$

has precisely one solution that is defined for all time $t$.

One could remedy this defect by multiplying the right-side of (1) by an appropriate scalar-valued function. For example, the differential equation

$$x' = \frac{1}{1 + |f(x)|} f(x) \tag{2}$$

does have the global existence property and the solution curves of (1) agree with those of (2). This is, of course, equivalent to introducing a new time parameter.

This change of time parameter has two defects. First, it can destroy certain dynamical properties that depend on a specific time parameter. Secondly, it has no obvious counterpart for nonautonomous differential equations.

It seems then that it is more appropriate to modify the definition of a flow so that a motion $\pi(x, t)$ does not have to be defined for all time $t$.

II.C. **Modified Definition of a Flow.** Let $X$ be a uniform space. For each point $x$ in $X$ let $I_x = (\alpha_x, \beta_x)$ be an open interval in $T$ containing 0. We shall assume that the intervals $I_x$ have the following continuity property: If $x_n \to x$ in $X$, then $I_x \subset \liminf I_{x_n}$. Let $D \subset X \times T$ be defined by

$$D = \{(x, t) \in X \times T : t \in I_x\}.$$

A function $\pi : D \to X$ is said to be *(local) flow on* $X$ if the following properties hold:

(i) (Identity property) $\pi(x, 0) = x$ for all $x$ in $X$;

(ii) (Group property) If $t \in I_x$ and $s \in I_{\pi(x,t)}$, then $(t + s) \in I_x$ and $\pi(\pi(x, t), s) = \pi(x, t + s)$;

(iii) $\pi$ is continuous;

(iv) (Maximal property) Each interval $I_x$ is maximal in the sense that either $I_x = T$, or the set

$$\{\pi(x, t) : 0 \leq t < \beta_x\}$$

is not conditionally compact if $\beta_x < +\infty$, or the set

$$\{\pi(x, t) : \alpha_x < t \leq 0\}$$

is not conditionally compact if $-\infty < \alpha_x$;

(v) The intervals $I_x$ are lower semi-continuous in $x$, that is if $x_n \to x$, then $I_x \subset \liminf I_{x_n}$.

It should be clear from the maximal property that if $E \subset D$ and $\pi : E \to X$ is a local flow, then $E = D$. Because of this there is no reason to distinguish between a local flow and a flow, and therefore we shall drop the modifier "local." Nevertheless, it will be convenient to have a way of referring to those flows that satisfy the first definition, viz,, those flows for which $I_x = T$ for all $x$; we shall call such a flow a **global flow.**

The following formulation of the continuity of $\pi$ will be needed in the sequel.

II.4. PROPOSITION. *Let $\pi$ be a flow on $X$. If $\{x_n\}$ is a (generalized) sequence in $X$ with $x_n \to x$, then the sequence of functions $\{\pi(x_n, t)\}$ converge to $\pi(x, t)$, and the convergence is uniform on compact sets in $I_x$.*

*Proof:* Let $K$ be a compact set in $I_x$. By the semicontinuity of $I_x$ we see that there is an index $m$ such that $K \subset I_{x_n}$ for $n \geq m$. The uniform convergence on $K$ means that for every neighborhood $V_a(\cdot)$, there is an index $p$ such that $\pi(x_n, t) \in V_a(\pi(x, t))$ for all $n \geq p$ and all $t$ in $K$. Assume, on the contrary, that this is false. Then there is a neighborhood $V_a(\cdot)$ such that for every index $n \geq m$ there is a $t_n$ in $K$ such that

$$\pi(x_n, t_n) \notin V_a(\pi(x, t_n)) . \tag{3}$$

We use the fact that since $K$ is compact, there is a convergent subsequence of $\{t_n\}$, which we shall denote by $\{t_n\}$. Say that $t_n \to t_0$. One then has $\pi(x, t_n) \to \pi(x, t_0)$.

With $a \in A$ given by (3) choose $b \in A$ by Proposition I.4 so that one has

$$\pi(x, t_0) \in V_b(\pi(x, t_n)) \text{ and } z \in V_b(\pi(x, t_0)) \Longrightarrow z \in V_a(\pi(x, t_n)). \tag{4}$$

It follows from the continuity of $\pi$, that $\pi(x_n, t_n) \to \pi(x, t_0)$. This means that we can find an index $N$ so that

$$z = \pi(x_n, t_n) \in V_b(\pi(x, t_0))$$

for $n \geq N$. By applying (4) we get $\pi(x_n, t_n) \in V_a(\pi(x, t_n))$ and this contradicts (3). QED.

Let $\pi$ be a flow on $X$ and let $x \in X$ be fixed. The function of $t$, $\pi(x, t)$ is said to be the **motion through** x. The **trajectory through** x is the set

$$\gamma(x) = \{\pi(x, t) : t \in I_x\} .$$

(This is also referred to as the **orbit** through x.) The **positive semitrajectory** and the **negative semitrajectory** are defined by

$$\gamma^+(x) = \{\pi(x, t) : 0 \leq t < \beta_x\}$$

and

$$\gamma^-(x) = \{\pi(x, t) : \alpha_x < t \leq 0\} ,$$

respectively. The **hull** of a point x is defined as

$$H(x) = \mathrm{Cl}\, \gamma(x) ,$$

and the **positive** *and* **negative hulls** are defined by

$$H^+(x) = \mathrm{Cl}\, \gamma^+(x) \text{ and } H^-(x) = \mathrm{Cl}\, \gamma^-(x) ,$$

respectively.

A set $E \subset X$ is said to be **invariant** if $\gamma(x) \subset E$ whenever $x \in E$. The set $E$ is said to be **positively invariant**, or **negatively invariant,** if $\gamma^+(x) \subset E$, or $\gamma^-(x) \subset E$, whenever $x \in E$.

It is easy to see that a set $E$ is invariant if and only if $E$ is the union of a collection of full trajectories.

II.5. PROPOSITION. *The closure of an invariant set is an invariant set.*

*Proof:* Let $E$ be an invariant set and let $y \in \mathrm{Cl}\, E$. We want to show that $\gamma(y) \subset \mathrm{Cl}\, E$, that is $\pi(y, t) \in \mathrm{Cl}\, E$ for each $t$ in $I_y$. Since $y \in \mathrm{Cl}\, E$ there is a (generalized) sequence $\{y_n\} \subset E$ such that $y_n \to y$. If $t \in I_y$, then by the continuity property of the intervals $I_x$, there is an index $m$ such that $t \in I_{y_n}$ for $n \geq m$. By the continuity of $\pi$ one has $\pi(y_n, t) \to \pi(y, t)$. Since $E$ is invariant, the subsequence $\{\pi(y_n, t): n \geq m\}$ is in $E$, hence $\pi(y, t) \in \text{Cl } E$. QED.

By the same argument we also see that the closure of a positively invariant set is positively invariant, and the closure of a negatively invariant set is negatively invariant.

Perhaps a "principle" should be noted here. It is apparent that there is a duality between the positive and negative properties of a flow. A statement valid for positive trajectories has an obvious counterpart for negative trajectories. In the sequel we shall examine the behavior for positive $t$ and leave unstated, but noted, the corresponding facts for negative time.

II.D. **Limit Sets and Compact Motions.** Let $\pi$ be a flow on $X$. Define the sets

$$LB^+ = \{x \in X: \beta_x = +\infty\}$$

$$LB^- = \{x \in X: \alpha_x = -\infty\}$$

$$LB = LB^- \cap LB^+ .$$

We note that if $LB$ is nonempty, then $LB \times T \subset D$, the domain of definition of $\pi$, and $\pi$, restricted to $LB \times T$, is a global flow on $LB$.

If $x \in LB^+$ we define the $\omega$-**limit set** by

$$\Omega_x = \cap_{t \geq 0} H^+(\pi(x, t)).$$

Similarly, if $x \in LB^-$ we define the $\alpha$-**limit set** by

$$A_x = \cap_{t \leq 0} H^-(\pi(x, t)) .$$

II.6. PROPOSITION. *The limit sets are described equivalently by*

$$\Omega_x = \{y \in X : y = \lim \pi(x, t_n) \text{ for some sequence } \{t_n\} \text{ with } t_n \to +\infty\}$$

and

$$A_x = \{y \in X : y = \lim \pi(x, t_n) \text{ for some sequence } \{t_n\} \text{ with } t_n \to -\infty\}.$$

*Proof:* By the duality principle it will suffice to prove the equality for the $\omega$-limit set. The sequences noted above are, of course, generalized sequences.

If $y = \lim \pi(x, t_n)$ for some sequence $\{t_n\}$ with $t_n \to +\infty$, then $y \in \mathrm{Cl}\,\gamma^+(x) = H^+(x)$. However,

$$y = \lim \pi(x, t_n) = \lim \pi(\pi(x, \tau), -\tau + t_n)$$

where $(-\tau + t_n) \to +\infty$. Hence $y \in H^+(\pi(x, \tau))$ for every $\tau \geq 0$. Therefor $y \in \Omega_x$.

Conversely, if $y \in \Omega_x$, then $y \in \mathrm{Cl}\,\gamma^+(\pi(x, \tau))$ for each $\tau \geq 0$. It follows from Proposition I.6 that there is a sequence $x_a$, defined on $A$ with range in $\gamma^+(\pi(x, \tau))$, such that $x_a \to y$. We can write this as $x_a = \pi(x, \tau + t_a)$ for some $t_a \geq 0$. With the usual ordering on $T^+$ we define the product ordering on $T^+ \times A$, that is $(\tau, a) \geq (\sigma, b)$ whenever $\tau \geq \sigma$ and $a \geq b$. We define a sequence $t$ on $T^+ \times A$ with range in $T^+$ by

$$t : n = (\tau, a) \to \tau + t_a = t_n .$$

First we note that $t_n \to +\infty$. To prove this we pick an integer $N$ and fix an index $(\sigma, b)$ in $T^+ \times A$ with $\sigma \geq N$. It follows that if $n = (\tau, a) \geq (\sigma, b)$ then

$$t_n = \tau + t_a \geq \sigma \geq N .$$

Hence $t_n \to +\infty$. Secondly we note that $y = \lim \pi(x, t_n)$, which is a direct consequence of Proposition I.3 on iterated limits. QED.

II.7. THEOREM. *The $\alpha$- and $\omega$-limit sets are closed and invariant.*

*Proof:* Consider the $\omega$-limit set $\Omega_x$. Since $\Omega_x$ is defined as the intersection of a family of closed sets, $\Omega_x$ is closed. If $y \in \Omega_x$, then there is a sequence $\{t_n\}$ with $t_n \to +\infty$ and $x_n = \pi(x, t_n) \to y$. By the continuity of $\pi$ for any $t$ in $I_y$ one has $\pi(x_n, t) \to \pi(y, t)$. However, $\pi(x_n, t) = \pi(x, t + t_n)$. Since $(t + t_n) \to +\infty$, it follows that $\pi(y, t) \in \Omega_x$ from the last proposition. Hence $\Omega_x$ is invariant. QED.

It should be noted here that the limit sets may be empty. This will be shown in the examples below. We would like to find a sufficient condition that the limit sets be nonempty. For this we introduce the following concept.

A motion $\pi(x, t)$ is said to be **compact** if the trajectory $\gamma(x)$ lies in a compact set, that is $\pi(x, t)$ is compact if the hull $H(x)$ is a compact set. Similarly a motion $\pi(x, t)$ is **positively compact** if $\gamma^+(x)$ lies in a compact set, that is if the positive hull $H^+(x)$ is a compact set. **Negative compactness** is defined similarly.

Because of the maximality of $I_x$ we see that if $\pi(x, t)$ is positively compact, then $x \in LB^+$. Similarly if $\pi(x, t)$ is negatively compact {compact}, then $x \in LB^-$ $\{x \in LB\}$. The following result shows the relationship between positively compact motions and their $\omega$-limit sets.

II.8. THEOREM. *Let $\pi(x, t)$ be a positively compact motion. Then $\Omega_x$ is nonempty, compact and invariant. Moreover, for every* *y in $\Omega_x$ one has $I_y = T$. If, in addition, the group $T$ is $R$, then the set $\Omega_x$ is connected.*

*Proof:* We have already observed that $\Omega_x$ is invariant. If $\pi(x, t)$ is positively compact, then $H^+(x)$ is a compact set in $X$. For $\tau \geq 0$, one has $H^+(\pi(x, \tau)) \subset H^+(x)$, hence $H^+(\pi(x, \tau))$ is a compact set in $X$. Since the family of sets $\{H^+(x, \tau)\}$ is decreasing and since the intersection of a decreasing family of nonempty compact sets is nonempty and compact, it follows that $\Omega_x$ is nonempty and compact.

If $y \in \Omega_x$, then $\pi(y, t)$ lies in the compact set $\Omega_x$ for all $t$ in $I_y$. Therefore from the maximality of $I_y$ we have $I_y = T$.

Now assume that $T = R$ and that $\Omega_x$ is not connected. Then there exist disjoint, nonempty, open sets $A$ and $B$ such that $\Omega_x \subset A \cup B$. We then can find (ordinary) sequences $\{t_n\}$ and $\{s_n\}$ so that

$$0 < s_1 < t_1 < \cdots < s_n < t_n < s_{n+1} < \cdots, \qquad s_n \to \infty, \quad t_n \to \infty$$

$$\pi(x, s_n) \in A \text{ and } \pi(x, t_n) \in B.$$

Since the path $\pi(x, [s_n, t_n])$ is connected it follows that there is a point $t_n^*$ in $[s_n, t_n]$ such that $\pi(x, t_n^*) \notin A \cup B$. From the compactness of $\pi(x, t)$, it follows that the sequence $\{\pi(x, t_n^*)\}$ contains a convergent subsequence, say that $\pi(x, t_n^*) \to y$. The limit point $y$ lies in the closed set $X - (A \cup B)$ and since $t_n^* \to +\infty$ one has $y \in \Omega_x$. This is a contradiction and so $\Omega_x$ is connected. QED.

II.9. COROLLARY. *Let $\pi$ be a flow on $X$. If there exists a positively compact motion, then the set $LB$ is nonempty.*

*Proof:* It follows from the last theorem that $\Omega_x$ is nonempty and lies in $LB$. QED.

II.E. **Minimal Sets.** A set $E \subset X$ is said to be a **minimal set** if it is nonempty, closed and invariant and if it contains no proper subset with these three properties. We shall undertake a detailed examination of minimal sets later in these lectures. For the moment we shall look at a number of elementary examples and a number of elementary properties.

A point $x$ in $X$ is said to be a **rest point (equilibrium point)** for the flow $\pi$ if $\pi(x, t) = x$ for all $t$. If $x$ is a rest point, then $E = \{x\}$ is a minimal set.

A point $x$ is said to be a **periodic point,** and the motion $\pi(x, t)$ is said to be **periodic,** if there is a $\tau > 0$ such that $\pi(x, \tau + t) = \pi(x, t)$ for all $t$. In this case we say that the motion $\pi(x, t)$ is $\tau$-**periodic.** The number $\tau$ is said to be a **period** of the motion $\pi(x, t)$. If $x$ is a periodic point, then $E = \gamma(x)$ is a minimal set. We shall see later that if $\pi(x, t)$ is compact, then $\gamma(x)$ is a minimal set if and only if $\pi(x, t)$ is periodic.

It should be noted that a periodic point may be a rest point. If it is not then one can show that every period is a multiple of a miniimal period. We leave the proof of the following statement as an exercise.

II.10. PROPOSITION. *Let $x$ be a periodic point and define $\sigma$ by*

$$\sigma = \inf\{\tau > 0 : \pi(x, \tau) = x\} .$$

*Then $x$ is a rest point if and only if $\sigma = 0$. Furthermore, if $\sigma > 0$, then any period $\tau$, of the motion $\pi(x, t)$, is an integral multiple of $\sigma$.*

In a later chapter we shall give other examples of minimal sets. The existence of minimal sets, or better, compact minimal sets is assured by the following result.

II.11. THEOREM. *If a flow* $\pi$ *admits a nonempty, compact, invariant set, then the flow has a compact minimal set. In particular, if there exists a positively compact motion* $\pi(x, t)$, *then the limit set* $\Omega_x$ *contains a compact minimal set.*

*Proof:* This is a simple application of Zorn's lemma, cf. [15]. Let $\mathcal{E} = \{E\}$ denote the collection of all nonempty, compact, invariant sets. Order $\mathcal{E}$ by set inclusion. If $\{E_\alpha\}$ is a chain in $\mathcal{E}$, that is $E_\alpha \subset E_\beta$ whenever $\beta \leq \alpha$ (the indices denote ordinal numbers), then the intersection $E = \bigcap_\alpha E_\alpha$ is an element of $\mathcal{E}$. Thus every chain in $\mathcal{E}$ has a lower bound, and we conclude that $\mathcal{E}$ has a minimal element $E_0$ It follows then that $E_0$ is a minimal set for the flow $\pi$.

If we apply the same argument to the collection $\mathcal{F} = \{E \cap \Omega_x :$ $E \in \mathcal{E}\}$, when $\pi(x, t)$ is a positively compact motion, we conclude that the minimal set $E_0$ can be chosen as a subset of $\Omega_x$. QED.

II.F. **Semi-Flows, Revisited.** We define a (local) **semi-flow** by the same properties that define a (local) flow but now restricting $t$ to be in the semigroup $T^+$. In other words, the intervals $I_x$ now are of the form $I_x = [0, \beta_x)$ where $\beta_x > 0$. Because of the maximality of the intervals $I_x$ we can, and do, drop the modifier "local" from this term.

The observations made for flows are also valid for semi-flows, with the appropriate modifications since the sets $\gamma^-(x)$ and $LB^-$ are not defined in a semi-flow. For example, if $\pi(x, t)$ is a positively compact motion in a semi-flow on $X$, then $\Omega_x$ is a nonempty, compact, *positively* invariant set in $X$.

II.G. **Attractors.** Let $\pi$ be a flow on a uniform space $X$ and let $M$ be a closed invariant subset of $X$. Assume that the restriction of $\pi$ to $M$ is a global flow. $M$ is said to be an **attractor** if there is an open set $U_0$ such that $M \subset U_0$ and for each $x \in U_0$

(i) $\pi(x, t) \in U_0$ for all $t \geq 0$;

(ii) $\pi(x, t) \to M$ as $t \to \infty$, that is, for every open neighborhood $V$ of $M$ there is a $\tau$ such that $\pi(x, t) \in V$ for all $t \geq \tau$.

An attractor $M$ is said to be **stable** if for every neighborhood $U$ of $M$ there is a neighborhood $V$ of $M$ such that $\pi(V, t) \subset U$ for all $t \geq 0$. If $M$ is an attractor we shall let $A(M)$ denote the **region of attraction** *of* $M$, that is $A(M)$ is the largest open set satisfying (i) and (ii) above.

A rather complete discussion of attractors can be found in [1]. We will need one important fact in the sequel.

II.12. LEMMA. *Let $M$ be a compact attractor for a flow $\pi$ on a uniform space $X$. If $x \in X$ has the property that $\gamma^+(x) \cap A(M) \neq \emptyset$, then $\Omega_x \subset M$. If in addition the space $X$ is locally compact, then $\Omega_x$ is nonempty.*

*Proof:* Assume that $y = \pi(x, \tau) \in A(M)$ for some $\tau \geq 0$. Then $\pi(y, t) = \pi(x, \tau + t) \to M$. This means that if $z \in \Omega_x = \Omega_y$, then $z = \lim \pi(x, \tau + t_n)$ for some sequence $\{t_n\}$ with $t_n \to \infty$, by Proposition II.6. Since the sequence $\{\pi(x, \tau + t_n)\}$ is eventually in every neighborhood of $z$ and eventually in every neighborhood of $M$, it follows from the Hausdorff property that $z \in M$.

If $X$ is locally compact, then there is a conditionally compact neighborhood $V$ of $M$ such that $M \subset V \subset \overline{V} \subset A(M)$. It follows from (ii) that $\pi(x, t) \in \overline{V}$ for $t \geq \tau$, hence $\pi(x, t)$ is positively compact. Hence $\Omega_x$ is nonempty by Theorem II.8. QED.

*Exercises.* 1. Define $F$: $R \to R$ by $Fx = -x$. Show that $F$ is not generated by a continuous flow on $R$.

2. Let $\pi: X \times R \to X$ be a continuous flow on $X$ and define $F$ by $F(x) = \pi(x, \tau)$ where $\tau$ is a fixed element of $R$. Show that $F$ is generated by a continuous flow on $X$.

3. Let $y = y(x)$ be a diffeomorphism of $R$ onto $R$. Find a necessary and sufficient condition that $y$ be generated by a continuous flow. Can the condition that $y(x)$ be a diffeomorphism be replaced by the condition that $y(x)$ be a homeomorphism?

4. Let $\pi$ be a flow on $X$ and assume that the set $LB$ is non-empty. Show that if $X$ is homeomorphic with a complete metric space, then $LB$ is homeomorphic with a complete metric space. (Hint: See Hausdorff's Theorem in [14].)

5. Prove the following converse of Proposition II.4: Let $X$ be a uniform space and let $\pi: X \times R \to X$ be a mapping. Show that $\pi$ is (jointly) continuous if and only if the following hold:

(i) If $x_n \to x$, then $\pi(x_n, t) \to \pi(x, t)$ uniformly for $t$ in compact sets.

(ii) If $t_n \to t$, then $\pi(x, t_n) \to \pi(x, t)$.

# CHAPTER III

## *EXAMPLES OF FLOWS AND SEMI-FLOWS*

In this chapter we shall consider a rather large collection of examples of flows and semi-flows. Much of the material here, in particular Examples A, B, E and I will be used later in these lectures. Other examples are presented to illustrate some refinements of the theory and they can be omitted by those lacking the background or the interest. However, most of the examples have been chosen because of their importance in applications, in particular applications to differential equations and Volterra integral equations

III.A. **Autonomous differential equations.** Let $v(x)$ be a $C^1$-vector field on an open set $W$ in $R^n$ and consider the ordinary differential equation

(1) $$x' = dx/dt = v(x) .$$

Let $\phi(x, t)$ be the solution of (1) that satisfies $\phi(x, 0) = x$. Then $\phi$ is a flow on $W$. If one only assumes $v(x)$ to be continuous and that (1) has unique solutions, then $\phi$ is still a flow on $W$.

It is possible to replace $W$ with a smooth manifold $M^n$, where $v(x)$ is a $C^1$-vector field on $M^n$. If $\phi$ is defined as above, then $\phi$ is a flow on $M^n$. It is worth noting that if the manifold $M^n$ is compact, then $\phi$ defines a global flow.

III.B. **Nonautonomous differential equations.** Let $v : W \times R \to R^n$ be a $C^1$-vector field, where $W$ is an open set in $R^n$, and consider the ordinary differential equation

(2) $$x' = v(x, t) .$$

Let $\phi(x_0, t_0, t)$ be the solution of (2) that satisfies $\phi(x_0, t_0, t_0) = x_0$ and let $I_{(x_0, t_0)}$ be the interval of definition of this solution. Define

$$J_{(x_0, t_0)} = \{t : t_0 + t \in I_{(x_0, t_0)}\} ,$$

and let $X = W \times R$. Define $\pi$ formally by

$$\pi(p, t) = (\phi(x_p, t_p, t_p + t), t_p + t)$$

where $p = (x_p, t_p) \in X$. The mapping $\pi$ defines a flow on $X$ where the interval of definition of the motion $\pi(p, t)$ is $J_p$.

This construction is equivalent to reducing Example B to Example A by writing (2) as

(2′) $$x' = v(x, t), \quad t' = 1 .$$

It should be noted that the flow $\pi$ for this case has no rest points, no periodic points, no compact motions and all the limit sets are empty. One of the objectives of these lectures is to overcome these deficiencies by giving a more appropriate definition of a flow for (2).

III.C. **Parametric equations.** Modify Example A so that the vector field $v$ depends continuously on a parameter $\mu$, where $\mu$ ranges over a uniform space $M$. Let $\phi(x, \mu, t)$ be the solution of

(3) $$x' = v(x, \mu)$$

that satisfies $\phi(x, \mu, 0) = x$. Let $X = W \times M$ and define $\pi$ by

$$\pi(x, \mu, t) = (\phi(x, \mu, t), \mu) .$$

Then $\pi$ is a flow on $X$.

III.D. **Difference Equations.** Let $W$ be an open set in $R^n$ and $f: W \to R^n$ a continuous function. Then the difference equation

$$u_{n+1} = f(u_n) \tag{4}$$

defines a discrete semi-flow on $W$ as follows: Let $\phi(u, n)$ be the solution of (4) that satisfies $\phi(u, 0) = u$. Let $I_u$ be the maximal interval of definition of $\phi$, that is either $I_\mu = I^+$ or $I_u = \{0, 1, \ldots, N\}$ where $\phi(u, n) \in W$ for $0 \leq n \leq N$ and $f(\phi(u, N)) \notin W$. (The set $I_u$ could consist of the single point $\{0\}$.) Then $\phi$ is a discrete semi-flow on $W$.

Note that if the range of $f$ is in $W$, then $I_\mu = I^+$ for all $u$. Also, if $f$ is a homeomorphism of $W$ onto $W$, then $\phi$ can be "backed-up," and $\phi$ defines a global flow.

III.E. **Continuous Functions.** Let $C = C(R, R^n)$ denote the space of continuous functions defined on $R$ with values in $R^n$. Assume that $C$ has the topology of uniform convergence on compact sets. That is, a sequence $\{f_k\}$ in $C$ is said to converge to a limit $f$ if for every compact set $K \subset R$ the sequence of restrictions $\{f_K | K\}$ converges to $f | K$ uniformly.

Let us show that this topology is metrizable. We let $I_n = [-n, n]$ and let $|\cdot|$ denote any norm on $R^n$. Now define

$$\sigma_n(f, g) = \sup\{|f(\theta) - g(\theta)| : \theta \in I_n\}$$

$$\rho_n(f, g) = \min\{1, \sigma_n(f, g)\}$$

$$\rho(f, g) = \Sigma_{n=1}^{\infty} 2^{-n} \rho_n(f, g)$$

when $f, g \in C$. We claim that $\rho$ is a metric on $C$ and that $\rho(f_k, f) \to 0$

if and only if $\{f_k\}$ converges to $f$ uniformly on every compact set in $R$. To prove that $\rho$ is a metric we first note that

1. $\sigma_n(f, g) = \sigma_n(g, f)$
2. $\sigma_n(f, f) = 0$
3. $\sigma_n(f, g) \leq \sigma_n(f, h) + \sigma_n(h, g)$

where $f, g, h \in C$. In other words, $\sigma_n$ is a pseudo-metric. Similarly, $\rho_n$ satisfies the same three conditions so $\rho_n$ and $\sigma_n$ are equivalent, which means that for any sequence $\{f_k\}$ in $C$ and any $f$ in $C$ one has $\rho_n(f_k, f) \to 0$ if and only if $\sigma_n(f_k, f) \to 0$. Since $|\rho_n(f, g)| \leq 1$, it follows that the series defining $\rho$ converges absolutely for every pair $f, g$ in $C$. It follows then that $\rho$ satisfies three conditions (1., 2., 3.) given above. We need only prove that $\rho(f, g) = 0$ implies that $f = g$. However $\rho(f, g) = 0$ if and only if $\rho_n(f, g) = 0$ for every $n$. Hence $f(\theta) = g(\theta)$ on every set $I_n$. Since $\cup_n I_n = R$, it follows that $f = g$. Hence $\rho$ is a metric on $C$.

The next thing to show is that the metric topology generated by $\rho$ agrees with the topology of uniform convergence on compact sets. Let $K$ be a compact set in $R$. Then for some integer $n$ one has $K \subset I_n$. Now if $\rho(f_k, f) \to 0$ for some sequence $\{f_k\}$, then $\rho_n(f_k, f) \to 0$, which implies that $\sigma_n(f_k, f) \to 0$. Hence the sequence $\{f_k\}$ converges uniformly to $f$ on the compact sets $I_n$ and $K$.

Going the other way, assume that $\{f_k\}$ converges to $f$ in the topology of uniform convergence on compact sets. We then want to show that for every $\varepsilon > 0$ there is an index $K$ such that $\rho(f_k, f) \leq 2\varepsilon$ whenever $k \geq K$. Let $\varepsilon > 0$ be given. Then choose $N$ so that one has

$$\sum_{n=N+1}^{\infty} 2^{-n} \leq \varepsilon$$

Since $\{f_k\}$ converges to $f$ uniformly on compact sets it converges uniformly on $I_N$. Therefore an index $K$ can be found so that the pseudo-metrics

$$\sigma_1(f_k, f), \ldots, \sigma_N(f_k, f)$$

are small whenever $k \geq K$. We choose $K$ so that

$$\sum_{n=1}^{N} 2^{-n} \rho_n(f_k, f) \leq \varepsilon\,, \qquad (k \geq K).$$

Then one has

$$\rho(f_k, f) \leq \sum_{n=1}^{N} 2^{-n} \rho_n(f_k, f) + \sum_{n=N+1}^{\infty} 2^{-n} \leq 2\varepsilon$$

whenever $k \geq K$.

We now define a mapping $\pi: C \times R \to C$ by

$$\pi(f, t) = f_t$$

where $f_t(\theta) = f(t + \theta)$. Let us show now that $\pi$ is a flow on $C$.

III.1 THEOREM. *The mapping* $\pi(f, t) = f_t$ *defines a continuous flow on* $C(R, R^n)$.

*Proof:* It is clear that $\pi(f, 0) = f$ and that $\pi$ satisfies the group property. Let us now show that $\pi$ is continuous. Let $\{f^{(k)}\}$ be a sequence in $C$ with limit $f$ and let $\{\tau_k\}$ be a sequence in $R$ with $\tau_k \to 0$. We want to show that for any $t$ in $R$ one has

$$\pi(f^{(k)}, t + \tau_k) \to \pi(f, t)\,.$$

There is no loss in generality in assuming that $|\tau_k| \leq 1$. Let $K$ be a compact set in $R$ and let $K_1$ be the compact set

$$K_1 = \{\theta + \tau:\ \theta \in K,\ |\tau| \leq 1\}\,.$$

Let $t$ in $R$ be fixed and let $\varepsilon > 0$ be given. Choose the index $N_1$ so that

$$|f_t^{(k)}(\theta') - f_t(\theta')| \leq \varepsilon , \qquad (\theta' \in K_1, \; k \geq N_1),$$

and choose $\delta > 0$, by the uniform continuity of $f_t$ on $K$, so that

$$|f_t(\theta + \tau) - f_t(\theta)| \leq \varepsilon , \qquad (\theta \in K, \; |\tau| \leq \delta).$$

Choose the index $N_2$ so that $|\tau_k| \leq \delta$ whenever $k \geq N_2$. Choose the index $N$ so that $N \geq N_1$ and $N \geq N_2$. Then for $k \geq N$ and $\theta \in K$ one has

$$|f_t^{(k)}(\theta + \tau_k) - f_t(\theta)| \leq |f_t^{(k)}(\theta + \tau_k) - f_t(\theta + \tau_k)|$$
$$+ |f_t(\theta + \tau_k) - f_t(\theta)| \leq 2\,\varepsilon ,$$

which establishes the continuity of $\pi$. QED.

It is clear that the mapping $(f, t) \to f_t$ defines a semi-flow on the subspace $C(R^+, R^n)$. We will use this fact in Chapter XI.

The dynamical system described above is sometimes referred to as the **Bebutov dynamical system.**

The following statement is easily verified.

III.2 PROPOSITION. *In the Bebutov dynamical system a point f is a rest point if and only if f is a constant function. A point f is a periodic point if and only if f is a periodic function.*

One may ask which motions $f_t$ are compact. The answer is given in the following statement.

III.3 THEOREM. *In the Bebutov dynamical system a motion* $f_t$ *is compact if and only if the function f is bounded and uniformly continuous.*

*Proof:* This is a direct application of Ascoli's Theorem, cf. [25], which states that a set $A$ in $C$ is conditionally compact if and only if

(i) $A$ is bounded, that is for each $\theta$ in $R$ there is a compact set $K$ in $R^n$ such that $f(\theta) \in K$ for all $f \in A$, and

(ii) $A$ is equicontinuous, that is for every $\theta$ in $R$ and every $\varepsilon > 0$ there is a $\delta = \delta(\varepsilon, \theta) > 0$ such that $|f(\theta) - f(\theta')| \leq \varepsilon$ for all $f$ in $A$, whenever $|\theta - \theta'| \leq \delta$.

Let us now prove the theorem. A motion $f_t$ is compact if and only if $\gamma(f)$ is conditionally compact. Now assume that $f_t$ is compact. Then $\gamma(f)$ is bounded, that is for $\theta = 0$ there is a number $B$ such that $|f_t(0)| = |f(t)| \leq B$ for all $t$ in $R$. Hence $f$ is bounded. Since $\gamma(f)$ is equicontinuous at $\theta_0 = 0$, for every $\varepsilon > 0$ there is a $\delta = \delta(\varepsilon) > 0$ such that for all $t$ in $R$ one has

$$|f_t(0) - f_t(\theta)| = |f(t) - f(t+\theta)| \leq \varepsilon$$

whenever $|\theta| \leq \delta$. Hence $f$ is uniformly continuous.

Going the other way, assume now that $f$ is bounded and uniformly continuous. It is clear then that $\gamma(f)$ is bounded. If $\delta(\varepsilon)$ is the modulus of uniform continuity for $f$, then

$$|f_t(\theta) - f_t(\theta')| = |f(t+\theta) - f(t+\theta')| \leq \varepsilon$$

for all $t$ in $R$ whenever $|\theta - \theta'| \leq \delta$. In other words, $\gamma(f)$ is equicontinuous. QED.

The above argument can be adapted to prove the next result.

III.4 THEOREM. *A motion $f_t$ in the Bebutov dynamical system is positively compact if and only if the function $f$ is bounded and uniformly continuous on $R^+$.*

Let us now look at a few examples of limit sets in this flow. For this purpose we shall assume that the functions are scalar-valued.

1. Let $f(\theta) = e^{-\theta}$. By Theorem III.4 the motion $f_t$ is positively compact ant $\Omega_f = \{0\}$, where 0 denotes the zero function.

2. Let $f(\theta) = e^{-\theta^2}$. In this case the motion $f_t$ is compact and $\Omega_f = A_f = \{0\}$.

3. Let $f(\theta) = \sin\theta$. Again the motion $f_t$ is compact and $\Omega_f = A_f = \gamma(f) = \{f_\tau : 0 \leq \tau < 2\pi\}$.

4. Let $f(\theta) = \sin\theta^2$. In this case $f$ is bounded but not uniformly continuous, and the limit sets $A_f$ and $\Omega_f$ are empty.

5. Let $f(\theta) = \sin\theta + \sin 2\pi\theta$. This is an example of a Bohr almost periodic function. We shall show later that $\Omega_f = A_f = H(f)$ and that a typical element $g$ in $\Omega_f$ is of the form

$$g(\theta) = \sin(\alpha + \theta) + \sin(\beta + 2\pi\theta)$$

where $0 \leq \alpha < 2\pi$ and $0 \leq \beta < 2\pi$.

III.F. **Sequences.** If we replace $R$ with the group of integers $I$ in the last example, then $C(I, R^n)$ is the space of all (ordinary) sequences with values in $R^n$. The assertions and arguments of the last example are true in this case. The topology now becomes the topology of point-wise convergence that is $f^{(k)} \to f$ in $C(I, R^n)$ if and only if $f^{(k)}(\theta) \to f(\theta)$ for each $\theta$ in $I$. The mapping $\pi$ given by

$$\pi(f, n) = f_n$$

defines a discrete flow on $C(I, R^n)$. In this case, a motion $f_n$ is compact if and only if the sequence $f$ is bounded, that is for some real number $B$ one has $|f(\theta)| \leq B$ for all $\theta$ in $I$. Similarly the motion $f_n$ is positively compact if and only if the sequence is bounded for $\theta \geq 0$.

III.G. **Bounded Uniformly Continuous Functions.** Let $BC = BC(R, R^n)$ denote the space of bound, continuous functions $f$ defined in $R$ with values in $R^n$ and let $BUC = BUC(R, R^n)$ denote the subspace of bounded, uniformly continuous functions. $BC$ is a Banach space with the norm

$$\|f\|_\infty = \sup\{|f(\theta)| : \theta \in R\},$$

and $BUC$ is a closed linear subspace.

Again we are interested in the mapping $(f, t) \to f_t$. We claim that for $f$ fixed this mapping is continuous in $t$ if and only if $f \in BUC$. Since

$$\|f_{t+\tau} - f_t\|_\infty = \sup\{|f(\tau+\theta) - f(\theta)| : \theta \in R\}$$

it follows that $\|f_{t+\tau} - f_t\|_\infty \to 0$ as $\tau \to 0$ if and only if $f$ is uniformly continuous. It is easy now to establish the next result.

III.5 THEOREM. *The mapping* $\pi(f, t) = f_t$ *defines a continuous flow on* $BUC$.

The compact motions in this flow will be of particular interest. We shall show later that a motion $f_t$ in $BUC$ is compact if and only if the function $f$ is Bohr almost periodic.

III.H. **Bounded Sequences.** Let $BC(I, R^n)$ denote the space of bounded sequences from $I$ into $R^n$. This is the Banach space $\ell_\infty$ with the norm

$$\|f\|_\infty = \sup\{|f(\theta)| : \theta \in I\}.$$

The mapping $\pi(f, n) = f_n$ defines a discrete flow on $BC(I, R^n)$.

III.I. **Continuous Functions with Parameters.** Let $W$ be an open set in Euclidean space $R^m$ and let $C = C(W\times R, R^n)$ denote the space of continuous functions $f$ defined on $W\times R$ with values in $R^n$. We shall assume that $C$ has the topology of uniform convergence on compact sets. This means that a sequence $\{f^{(k)}\}$ in $C$ converges to a function $f$ in $C$ if and only if for every set $K\times I$, where $K$ is a compact set in $W$ and $I$ a compact set in $R$, the sequence $\{f^{(k)}\}$ converges to $f$ uniformly on $K\times I$. This topology is metrizable. Indeed, if we let $\{K_n\}$ be an increasing sequence of compact sets in $W$ with $W = \bigcup_n K_n$ and $I_n = [-n, n]$ and define

$$\sigma_n(f, g) = \sup\{|f(x, \theta) - g(x, \theta)| : x \in K_n, \theta \in I_n\}$$

and define $\rho_n$ and $\rho$ as in Example E, then $\rho$ is a metric on $C$. By using the argument of Example E, it becomes easy to verify that $\rho$ generates the topology of uniform convergence on compact sets.

We define a mapping $\pi : C\times R \to C$ by

$$\pi(f, t) = f_t$$

where $f_t(x, \theta) = f(x, t+\theta)$. This is a continuous flow on $C$, as we state in the next theorem. The proof of this result is very similar to the proof of Theorem III.1, and we leave the details as an exercise.

III.6 THEOREM. *The mapping* $\pi(f, t) = f_t$ *defines a continuous flow on* $C(W\times R, R^n)$.

It is interesting to ask when a point $f$ in $C$ is a rest point. The answer, which follows directly from Proposition III.2, is that $f$ is a rest point if and only if $f$ is independent of $t$. In later applications the functions $f$ from $C$ will arise as the right-side of differential equations of the form

$$x' = f(x, t) .$$

It follows then that $f$ is a rest point in $C$ if and only if the differential equation is autonomous.

Similarly we see that a point $f$ is a periodic point in $C$ if and only if $f(x, \theta)$ is periodic in $\theta$, This means that there is a $\tau > 0$ such that $f(x, \theta+\tau) = f(x, \theta)$ for all $(x, \theta)$ in $W \times R$.

The question of compact, or positively compact, motions in $C$ is settled by a straightforward modification of Theorems III.3 and III.4.

III.7 THEOREM. (A) *A motion $f_t$ in $C(W \times R, R^n)$ is compact if and only if the function $f$ is bounded and uniformly continuous on every set $K \times R$, where $K$ is a compact set in $W$.*

(B) *A motion $f_t$ in $C(W \times R, R^n)$ is positively compact if and only if the function $f$ is bounded and uniformly continuous on every set $K \times R^+$, where $K$ is a compact set in $W$.*

*Proof:* Again we shall use Ascoli's Theorem, cf. [25] and Theorem III.3. The motion $f_t$ is compact if and only if the trajectory $\gamma(f)$ is conditionally compact. If $\gamma(f)$ is conditionally compact, then $\gamma(f)$ is bounded; that is for every $(x, t)$ the set $\{f_\tau(x, t) : \tau \in R\}$ is conditionally compact in $R^n$. In other words, for each $x$ the function $f(x, t)$ is bounded in $t$. Also, the set $\gamma(f)$ is equicontinuous at the point $(x_0, t_0)$ if for every $\varepsilon > 0$ there is a $\delta = \delta(\varepsilon, x_0, t_0) > 0$ such that

$$(5) \qquad |f_\tau(x, t) - f_\tau(x_0, t_0)| = |f(x, \tau+t) - f(x_0, \tau+t_0)| < \varepsilon , \qquad (\tau \in R)$$

whenever $|x - x_0| < \delta$ and $|t - t_0| < \delta$. This shows that the $\delta$ can be chosen independent of $t_0$. Then (5) is equivalent to saying that $f$ is uniformly continuous on sets of the form $K \times R$ where $K$ is compact in $W$. Since this reasoning process can be reversed we have thus shown that $\gamma(f)$ is conditionally compact if and only if

(a) for each $x$, the function $f(x, t)$ is bounded in $t$, and

(b) the function $f$ is uniformly continuous on every set of the form $K \times R$, where $K$ is compact in $W$.

Finally, we note that this is equivalent to saying that $f$ is bounded and uniformly continuous on every set of the form $K \times R$, where $K$ is compact in $W$. This completes the proof of (A). The proof of statement (B) is similar. QED.

*Application:* In the study of perturbation problems in the theory of differential equations one often finds equations where the right-sides satisfy the following conditions. Let $g, h \in C(W \times R, R^n)$ and define $f$ by $f = g + h$. Assume that

$$(6) \qquad h(x, \theta) \to 0 \text{ as } \theta \to \infty$$

uniformly for $x$ in compact sets in $W$. One can then show that $\Omega_f = \Omega_g$, that is the "ultimate" behavior of $f$ and $g$ agree. To prove this we note that (6) is equivalent to saying that for every sequence $\{t_n\}$ in $R$ with $t_n \to +\infty$ one has $h_{t_n} \to 0$ in $C$. Since $f_t = g_t + h_t$ for every $t$, it follows from Proposition II.6 that $\Omega_f = \Omega_g$. For example, if $g(x, \theta)$ happened to be $\tau$-periodic in $\theta$, then $\Omega_f = \Omega_g = \{g_t: 0 \leq t \leq \tau\}$. It should also be clear that the motion $f_t$ is positively compact if and only if $g_t$ is positively compact.

III.J. **Sequences with parameters.** By replacing $R$ with the integers $I$ in the last example one defines a space $C(W \times I, R^n)$. The translation mapping

$$\pi(f, t) = f_t \qquad (t \in I)$$

defines a discrete flow on $C(W \times I, R^n)$. We invite the reader to derive the characterization of compact motions for this flow.

III.K **Markov chains.** Let $X$ denote the manifold in $R^n$ determined by the vectors $x = (x_1, \ldots, x_n)$ for which $x_i \geq 0$ $(1 \leq i \leq n)$ and $\Sigma_{i=1}^n x_i = 1$. Let $A = (a_{ij})$ be an $n \times n$ stochastic matrix, which means that $a_{ij} \geq 0$ and $\Sigma_{i=1}^n a_{ij} = 1$ $(1 \leq j \leq n)$. The matrix $A$ can then be viewed as the transition matrix for a Markov process acting on a set of $n$ states. The probability distribution $x$ in $X$ is mapped onto $y = Ax$ in $X$. The mapping

$$(7) \qquad \pi(x, n) = A^n x, \qquad (n = 0, 1, \ldots)$$

defines a discrete semi-flow on $X$. The Chapman- Kolmogorov equation for this Markov process is simply a formulation of the group property for $\pi$, cf. [11].

III.8 PROPOSITION. *The discrete semi-flow defined by* (7) *has at least one rest point.*

*Proof:* First we note that the rest points for $\pi$ are precisely the fixed points of the mapping $x \to Ax$. Since this mapping is a continuous mapping of $X$ into $X$, and since $X$ is homeomorphic with a $(n-1)$ disk, it follows from the Brouwer Fixed Point Theorem, cf. [7], that the mapping $x \to Ax$ has at least one fixed point in $X$. QED.

III.L **Measurable Functions.** For $1 \leq p < \infty$ let $\mathcal{L}_p(R, R^n)$ denote the space of all measurable functions $f$ defined on $R$, with values

c

in $R^n$, such that for any compact set $I \subset R$ the function $f$ belongs to the Lebesgue space $L_p(I, R^n)$, that is $\int_I |f(\theta)|^p d\theta < \infty$, where $\int$ denotes the Lebesgue integral. We define a metric on $\mathfrak{L}_p(R, R^n)$ as follows: First set

$$\sigma_n(f, g) = \left( \int_{-n}^{n} |f(\theta) - g(\theta)|^p d\theta \right)^{1/p}$$

and now define $\rho_n$ and $\rho$ as in Example E. One can show that $\rho$ is a metric and that $\rho(f^{(k)}, f) \to 0$ if and only if $f^{(k)} \to f$ in every Lebesgue space $L_p(I, R^n)$ where $I$ is a compact set in $R$, that is $\rho(f^{(k)}, f) \to 0$ if and only if

$$\int_I |f^{(k)}(\theta) - f(\theta)|^p d\theta \to 0$$

for every compact interval $I$.

We wish to show that the translation mapping

$$\pi(f, t) = f_t$$

defines a continuous flow on $\mathfrak{L}_p(R, R^n)$. The identity property and the group property are easily verified. The difficulty here, as is the case in most examples, is showing that $\pi$ is continuous. We shall first show that $\pi$ is continuous in each variable separately.

III.9 Lemma. *If $f^{(k)} \to f$ in $\mathfrak{L}_p(R, R^n)$, then one has $f_t^{(k)} \to f_t$ in $\mathfrak{L}_p(R, R^n)$ uniformly for $t$ in compact sets.*

The proof of this statement is left as an exercise.

III.10 Lemma. *Let $f \in \mathfrak{L}_p(R, R^n)$. Then the mapping $t \to f_t$ of $R$ into $\mathfrak{L}_p(R, R^n)$ is continuous.*

*Proof:* Let $K$ be a compact set in $R$ and let $t \in R$ be fixed. Let $\varepsilon > 0$ be given. We then want to show that there is a $\delta > 0$ such that

$$\int_K |f_{t+\tau}(\theta) - f_t(\theta)|^p \, d\theta \leq \varepsilon \tag{8}$$

provided $|\tau| \leq \delta$. We shall prove (8) in three steps.

(1) First we shall assume that $f$ is continuous,

(2) then we shall assume that $f$ is bounded on compact sets, and

(3) finally we shall treat the general case.

Case 1. Assume that $f$ is continuous. Then $f$ is uniformly continuous on compact sets. Hence there is a $\delta > 0$ such that

$$|f_t(\theta+\tau) - f_t(\theta)| \leq (\varepsilon/|K|)^{1/p}, \qquad (\theta \in K, \ |\tau| \leq \delta),$$

where $|K|$ denotes the measure of $K$. Inequality (8) now follows.

Case 2. Assume that $f$ is bounded on compact sets. We shall now show that $f$ can be approximated in $\mathcal{L}_p(R, R^n)$ by a continuous function. Let $h$, $0 < h \leq 1$, be given and define

$$\hat{f}(\theta) = \hat{f}^{(h)}(\theta) = \frac{1}{h} \int_\theta^{\theta+h} f(\xi) d\xi$$

Then $\hat{f}$ is continuous and we claim that $\hat{f} \to f$ in $\mathcal{L}_p(R, R^n)$ as $h \to 0$. To prove the convergence property we first note that

$$\hat{f}(\theta) \to f(\theta), \qquad \text{(almost everywhere)}$$

as $h \to 0$, cf. [9]. Secondly, since $f$ is bounded there is a number $B$ such that $|f(\theta')| \leq B$ on the compact set

$$K' = \{\theta' = \theta + h : \ \theta \in K, \ 0 \leq h \leq 1\}.$$

Therefore

$$|\hat{f}(\theta)|^p \leq \left[\frac{1}{h}\int_{\theta}^{\theta+h} |f(\xi)|\,d\xi\right]^p \leq B^p$$

on the compact set $K$. By the Lebesgue Dominated Convergence Theorem we have

$$\int_K |\hat{f}(\theta) - f(\theta)|^p\, d\theta \to 0, \quad \text{as } h \to 0,$$

and thus $\hat{f} \to f$ (as $h \to 0$) in $\mathfrak{L}_p(R, R^n)$. Also by Lemma III.9 we have $\hat{f}_t \to f_t$ in $\mathfrak{L}_p(R, R^n)$. Furthermore, the convergence

$$\hat{f}_{t+\tau} \to f_{t+\tau} \qquad (\text{as } h \to 0)$$

is uniform for $|\tau| \leq 1$. Now choose $h$ so small that

$$\int_K |\hat{f}_{t+\tau}(\theta) - f_{t+\tau}(\theta)|^p d\theta \leq (\varepsilon/3)^p \tag{9}$$

for $|\tau| \leq 1$. With $h$ so fixed, choose $\delta$ (by Case 1) so that $0 < \delta \leq 1$ and

$$\int_K |\hat{f}_{t+\tau}(\theta) - \hat{f}_t(\theta)|^p d\theta \leq (\varepsilon/2)^p \tag{10}$$

when $|\tau| \leq \delta$. From the Minkowski inequality, cf. [9], we get

$$\left(\int_K |f_{t+\tau}(\theta) - f_t(\theta)|^p d\theta\right)^{1/p} \leq \left(\int_K |f_{t+\tau}(\theta) - \hat{f}_{t+\tau}(\theta)|^p d\theta\right)^{1/p}$$

$$+\left(\int_K |\hat{f}_{t+\tau}(\theta) - \hat{f}_t(\theta)|^p d\theta\right)^{1/p} + \left(\int_K |\hat{f}_t(\theta) - f_t(\theta)|^p d\theta\right)^{1/p}$$

and by applying (9) and (10) we get (8) when $|\tau| \leq \delta$.

Case 3. If $f$ is not bounded, then we perform the usual truncation process and reduce this to Case 2. That is, if

$$f_i^{(m)}(\theta) = \begin{cases} m, & \text{if } f_i(\theta) > m \\ f_i(\theta), & \text{if } -m \leq f_i(\theta) \leq m \\ -m, & \text{if } f_i(\theta) < -m \end{cases}$$

where $f_i$ denotes the $i^{\text{th}}$ component of $f$, then $f^{(m)} \to f$ in $\mathcal{L}_p(R, R^n)$. We then choose $m$ sufficiently large and apply the result of Case 2 to the fixed $f^{(m)}$. The details are straightforward and we omit them. QED.

III.11 THEOREM. *The mapping* $\pi(f, t) = f_t$ *defines a continuous flow on* $\mathcal{L}_p(R, R^n)$.

*Proof:* As we have already observed, the identity property and the group property are easily verified, so it remains to show that $\pi$ is continuous. Let $\{f^{(k)}\}$ be a sequence in $\mathcal{L}_p(R, R^n)$ with $f^{(k)} \to f$ and let $\{\tau_k\}$ be a sequence in $R$ with $\tau_k \to 0$. We want to show that for any $t$ in $R$ one has

$$\pi(f^{(k)}, t+\tau_k) \to \pi(f, t)$$

in $\mathcal{L}_p(R, R^n)$. First choose an index $m$ so that $|\tau_k| \leq 1$ when $k \geq m$, and now restrict attention to the subsequences $\{f^{(k)}: k \geq m\}$ and $\{\tau_k : k \geq m\}$. For any compact set $K$ in $R$ one has

$$\left(\int_K |f_t^{(k)}(\theta+\tau_k) - f_t(\theta)|^p d\theta\right)^{1/p} \leq \left(\int_K |f_t^{(k)}(\theta+\tau_k) - f_t(\theta+\tau_k)|^p d\theta\right)^{1/p} + \left(\int_K |f_{t+\tau_k}(\theta) - f_t(\theta)|^p d\theta\right)^{1/p}$$

The first term on the right-side becomes small (as $f^{(k)} \to f$) by

Lemma III.9, and the second term becomes small (as $\tau_k \to 0$) by Lemma III.10. QED.

It is clear that the mapping $(f, t) \to f_t$ $(t \geq 0)$ defines a semi-flow on the subspace $\mathfrak{L}_p(R^+, R^n)$.

It is easy to characterize the rest points and periodic points in this flow. Compact motions are a bit harder to verify. We state the following characterization and refer the reader to [30] for a proof.

III.12 THEOREM. *A motion $f_t$ is compact {positively compact} in $\mathfrak{L}_p(R, R^n)$ if and only if*

(i) *there is a $B$ in $R$ such that $\int_0^1 |f(\theta+t)|^p d\theta \leq B$ for all $t$ in $R$ {$t$ in $R^+$}, and*

(ii) *for every $\varepsilon > 0$ there is a $\delta > 0$ such that*

$$\int_0^1 |f(\theta+t+h) - f(\theta+t)|^p d\theta \leq \varepsilon$$

*whenever $|h| \leq \delta$ and $t \in R$ {$t \in R^+$}.*

Let $f = g + h$ where $g$ and $h$ are in $\mathfrak{L}_p(R, R^n)$. Assume that $g$ is bounded and uniformly continuous on $R^+$ and that $\int_0^\infty |h(\theta)|^p d\theta < \infty$. Then the motion $f_t$ is positively compact in $\mathfrak{L}_p(R, R^n)$ and furthermore $\Omega_f = \Omega_g$. This fact and related results are discussed in the exercises at the end of the chapter.

III.M. **The space** $\mathfrak{L}_\infty(R, R^n)$. In the last example we discussed the flow generated by the translation mapping $f_t$ on the spaces $\mathfrak{L}_p(R, R^n)$ for $1 \leq p < \infty$. The space $\mathfrak{L}_\infty(R, R^n)$ is defined as the space of all measurable functions $f$ defined on $R$, with values in $R^n$, such that for any compact set $I \subset R$ the function $f$ belongs to the Lebesgue space $L_\infty(I, R^n)$, that is

$$\text{ess. sup } \{|f(\theta)| : \theta \in I\} < \infty .$$

One can construct a metric on $\mathcal{L}_\infty(R, R^n)$ by setting

$$\sigma_n(f, g) = \text{ess. sup}\{|f(\theta) - g(\theta)| : -n \leq \theta \leq n\}$$

and defining $\rho_n$ and $\rho$ as in Example E.

If we would now consider the translation mapping $\pi(f, t) = f_t$ in $\mathcal{L}_\infty(R, R^n)$ with the metric topology defined above, it would be easy to establish the counter-part of Lemma III.9, that is, the mapping $\pi(f, t) = f_t$ is continuous in the $f$-variable. However, continuity of $\pi$ in the $t$-variable is not valid in general. In fact it is easy to show that the mapping

$$\pi(f, \cdot) : R \to \mathcal{L}_\infty(R, R^n),$$

where $t \to f_t$, is continuous if and only if the function $f(\cdot)$ is continuous, that is $f \in C(R, R^n)$. (Compare this with the situation in Example G.)

One can define another topology on $\mathcal{L}_\infty(R, R^n)$, a topology which we shall call the *weak *-topology*. It is generated by the weak*-topology on $L_\infty(I, R^n)$ where we note that $L_1(I, R^n)^* = L_\infty(I, R^n)$. (See Example 3 in Chapter I.) We say that a (generalized) sequence $\{f^{(m)}\}$ in $\mathcal{L}_\infty(R, R^n)$ converges to a limit $f$ in $\mathcal{L}_\infty(R, R^n)$ if for every compact interval $I \subset R$ and every $x(\cdot)$ in $L_1(I, R^n)$ one has

$$\int_I [f^{(m)}(\theta) - f(\theta)] \cdot x(\theta)\, d\theta \to 0$$

where $\cdot$ denotes the usual dot product of vectors. This defines a topology which can be given a uniform structure as follows: The index set $A$ consists of elements of the form

$$a = (I\,;\ x_1, \ldots, x_m\,;\ \varepsilon)$$

where $I$ is a compact set in $R$, $x_i(\cdot) \in L_p(I, R^n)$ $(1 \leq i \leq m)$ and $\varepsilon > 0$ and the correspondence $V_a(f)$ is defined by

$$V_a(f) = \{g \in \mathfrak{L}_\infty(R, R^n) : \left| \int_I [f(\theta) - g(\theta)] \cdot x_i(\theta) d\theta \right| \leq \varepsilon^{-1},\ 1 \leq i \leq m\}.$$

The ordering on $A$ is defined by

$$a = (I\,;\ x_1, \ldots, x_m\,;\ \varepsilon) \leq b = (J\,;\ y_1, \ldots, y_k\,;\ \delta)$$

if and only if $I \subset J$, $\{x_1, \ldots, x_m\} \subset \{y_1, \ldots, y_k\}$ and $\varepsilon \leq \delta$.

In this topology the translation mapping is now jointly continuous and in fact we have:

III.13 THEOREM. *The mapping* $\pi(f, t) = f_t$ *defines a continuous flow on* $\mathfrak{L}_\infty(R, R^n)$, *where* $\mathfrak{L}_\infty(R, R^n)$ *has the weak*-topology.*

*Proof:* The argument follows that used in the proof of Theorem III.11. We shall show that the mapping $\pi$ is continuous in the $t$-variable and ask the reader to check the other details.

Let $\{\tau_m\}$ be a (generalized) sequence in $R$ with $\tau_m \to 0$. We want to show that for every $t$ in $R$ one has $f_{t+\tau_m} \to f_t$. This means that for every compact set $I \subset R$ and every $x(\cdot)$ in $L_1(I, R^n)$ one has

$$\int_I [g(\theta+\tau_m)-g(\theta)]\cdot x(\theta)\,d\theta \to 0$$

where $g = f_t$. However,

$$\left|\int_I [g(\theta+\tau_m)-g(\theta)]\cdot x(\theta)d\theta\right| \le \int_I |g(\theta+\tau_m)\,x(\theta+\tau_m)-g(\theta)\cdot x(\theta)|\,d\theta$$
$$+ \left|\int_I g(\theta+\tau_m)\cdot[\,x(\theta)-x(\theta+\tau_m)]d\theta\right|$$

The first term on the right-side becomes small (as $\tau_m \to 0$) by Lemma III.10 while noting that the function $h(\theta) = g(\theta)\cdot x(\theta)$ is in $\mathcal{L}_1(R, R)$. For the second term we first choose the index $M$ so that $|\tau_m| \le 1$ for $m \ge M$. Then by the Hölder inequality, cf.[9], we have for $m \ge M$

$$\left|\int_I g(\theta+\tau_m)\cdot[x(\theta)-x(\theta+\tau_m)]d\theta\right| \le B\cdot \int_I |x(\theta)-x(\theta+\tau_m)|\,d\theta$$

where

$$B = \text{ess sup}\{|g(\theta')| : \theta' = \theta+\tau,\ \theta \in I \text{ and } |\tau| \le 1\}.$$

Now by applying Lemma III.10 to the function $x(\cdot)$, the continuity assertion follows. QED.

Compact motions in this flow are particularly easy to characterize. We use the fact that the closed, bounded sets are compact in the weak*-topology, cf. [9].

III.14 THEOREM. *A motion $f_t$ in the flow on $\mathcal{L}_\infty(R, R^n)$, where $\mathcal{L}_\infty(R, R^n)$ has the weak*-topology, is compact if and only if*

$$\|f\|_{\infty} = \text{ess sup}\{|f(\theta)| : \theta \in R\} < \infty,$$

*that is,* $f \in L_{\infty}(R, R^n)$.

*Proof:* First assume that $f \in L_{\infty}(R, R^n)$ and now let $\{f^{(n)}\}$ be any (generalized) sequence in the trajectory $\gamma(f) = \pi(f, R)$. We want to show that there exists a convergent subsequence. Convergence now means convergence in the weak*-topology on each $L_{\infty}(I, R^n)$ where $I$ is compact. However,

$$\text{ess sup}\{|f^{(n)}(\theta)| : \theta \in I\} \leq \|f\|_{\infty} < \infty .$$

Since a bounded set in $L_{\infty}(I, R^n)$ is compact in the weak*-topology, it follows that we can find a subsequence of $\{f^{(n)}\}$ converging in $L_{\infty}(I, R^n)$. By taking a family of such intervals $\{I_k\}$ with $I_k \subset I_{k+1}$ and $R = \cup_k I_k$, and by using the usual diagonalization argument one constructs a subsequence of $\{f^{(n)}\}$ convergent in $\mathfrak{L}_{\infty}(R, R^n)$.

Now assume that the motion $f_t$ is compact in $\mathfrak{L}_{\infty}(R, R^n)$. Then the family of functions $\{f_t(\cdot) : t \in R\}$ forms a conditionally compact set in $L_{\infty}([0, 1], R^n)$. Hence it lies in a bounded set in $L_{\infty}([0, 1], R^n)$, that is, there is a $B < \infty$ such that

$$|f_t(\theta)| = |f(t+\theta)| \leq B, \qquad (0 \leq \theta \leq 1, \ t \in R).$$

This implies that $\|f\|_{\infty} \leq B$, or that $f \in L_{\infty}(R, R^n)$. QED.

III.N. **Measurable Functions with Parameters.** In this paragraph we shall generalize Example L in the same way that Example I generalizes the Bebutov dynamical system. Let $p$ be given, where $1 \leq p < \infty$, and define $\mathfrak{L}_p(W \times R, R^n)$ to be the collection of all measurable functions $g(x, t)$ defined on $W \times R$ with range in $R^n$ such that for each compact set $K \subset W$ there are measurable functions $m(t)$ and $k(t)$ satisfying

$$|g(x,t)] \leq m(t), \qquad (x \in K,\ t \in R),$$

$$|g(x,t)-g(y,t)| \leq k(t)\,|x-y| \qquad (x, y \in K,\ t \in R),$$

where $m$ and $k$ are locally $L_p$.

We will be interested in topologies on $\mathfrak{L}_p(W \times R, R^n)$ under which the translation mapping $\pi(g, t) = g_t$ defines a continuous flow. Two topologies are considered. We shall denote them by $\mathfrak{T}_b$ and $\mathfrak{T}_c$. These topologies arise in the study of differential equations (cf. [23]) and Volterra integral equations (cf. [19],[20]).

We shall say that a (generalized) sequence $\{g_n\}$ in $\mathfrak{L}_p(W \times R, R^n)$ **converges (in $\mathfrak{T}_c$)** to a limit $g$ if for every compact interval $I \subset R$ and every compact set $\mathcal{K} \subset C(I, W)$ one has

$$\sup_{x(\cdot) \in \mathcal{K}} \int_I |g_n(x(s), s) - g(x(s), s)|^p ds \to 0$$

(The subscript of $\mathfrak{T}_c$ refers to the fact that $\mathcal{K}$ is a "compact" set in $C(I, W)$, or in $C(I, R^n)$.) We shall say that the (generalized) sequence $\{g_n\}$ **converges ( in $\mathfrak{T}_b$)** to a limit $g$ if for every compact interval $I \subset R$ and every compact set $K \subset W$ one has

$$\sup_{x(\cdot) \in \mathcal{K}} \int_I |g_n(x(s), s) - g(x(s), s)|^p ds \to 0$$

where $\mathcal{K} = C(I, K)$. (The subscript of $\mathfrak{T}_b$ refers to the fact that $\mathcal{K} = C(I, K)$ is a "bounded" set in $C(I, R^n)$. ) We shall use the notation $(\mathfrak{L}_p, \mathfrak{T}_c)$ and $(\mathfrak{L}_p, \mathfrak{T}_b)$ to refer to $\mathfrak{L}_p(W \times R, R^n)$ with the respective topologies. Both of these topologies are uniform topologies and we ask the reader to verify this fact. In fact the topology $\mathfrak{T}_b$ is metrizable.

First consider $(\mathfrak{L}_p, \mathfrak{T}_b)$. We shall now prove:

III.15 THEOREM. *The mapping* $\pi(g, t) = g_t$ *defines a continuous flow on* $(\mathfrak{L}_p, \mathfrak{T}_b)$.

The proof of this follows the argument used in the proof of Theorem III.11. That is, first one shows that the mapping $\pi(g, t) = g_t$ is continuous in each variable separately, and then one shows joint continuity. The identity property and the group property for flows present no problem in this case.

III.16 LEMMA. *If* $g^{(k)} \to g$ *in* $(\mathfrak{L}_p, \mathfrak{T}_c)$, *or* $(\mathfrak{L}_p, \mathfrak{T}_b)$, *then* $g_t^{(k)} \to g_t$ *in* $(\mathfrak{L}_p, \mathfrak{T}_b)$, *uniformly for* $t$ *in compact sets.*

The proof of this statement is left as an exercise.

III.17. LEMMA. *Let* $g \in \mathfrak{L}_p(W \times R, R^n)$. *Then the mapping* $t \to g_t$ *of* $R$ *into* $(\mathfrak{L}_p, \mathfrak{T}_b)$, *is continuous.*

*Proof:* The proof of this follows the argument used in Lemma III.10. The only major difference occurs in showing that the function $g$ can be approximated by continuous functions in $(\mathfrak{L}_p, \mathfrak{T}_b)$. We shall prove this fact here and ask the reader to verify the other details. It is this aspect of the argument that shows the reason for the Lipschitz condition on $g$, a condition that was not needed for the flow on $C(W \times R, R^n)$ in Example I.

Assume that $g(x, t)$ is bounded on compact sets and let $K$ be a compact set in $W$ and $I = [a, b]$ a compact interval in $R$. Let $J = [a, b+1]$ and define $\hat{g}(x, t)$ by

$$\hat{g}(x, t) = \frac{1}{h} \int_t^{t+h} g(x, s)\,ds$$

where $0 < h \leq 1$. We shall show that $\hat{g} \to g$ in $(\mathfrak{L}_p, \mathfrak{T}_b)$ as $h \to 0$.

Define $F(x, t, h)$ by

$$F(x, t, h) = g(x, t) - \frac{1}{h} \int_t^{t+h} g(x, s)ds.$$

If $|g(x, t)| \leq B$ on the compact set $K \times J$, then for any $h$, $0 < h \leq 1$, and any continuous function $x(\cdot)$ in $\mathcal{K} = C(J, K)$ one has

$$|F(x(t), t, h)| \leq |g(x(t), t)| + \frac{1}{h} \int_t^{t+h} |g(x(s), s)| ds \leq 2B.$$

Since $F(x(t), t, h) \to 0$, almost everywhere, as $h \to 0$ (cf. [9]), it follows from the Lebesgue Dominated Convergence Theorem that

$$\int_I |F(x(t), t, h)|^p dt \to 0 \qquad \text{(as } h \to 0) \tag{11}$$

for each $x(\cdot)$ in $\mathcal{K}$.

We want to show that the convergence in (11) is uniform for $x(\cdot)$ in $\mathcal{K}$. To prove this we note that $g(x, t)$ is locally Lipschitzian in $x$. Let $k(t)$ be the Lipschitz constant for $g$ on the compact set $K$. Then

$$\hat{k}(t) = \frac{1}{h} \int_t^{t+h} k(s)ds$$

is the Lipschitz constant for $\hat{g}$ on $K$ and

$$|F(x, t, h) - F(y, t, h)| \leq [k(t) + \hat{k}(t)]|x - y|$$

for $x, y \in K$, $0 < h \leq 1$ and $t \in R$. Since $\hat{k} \to k$ in $\mathfrak{L}_p(R, R)$ (cf. proof of Lemma III.10), then for $h$ sufficiently small, say $0 < h \leq h_0$, one has

$$\left[\int_I \hat{k}(t)^p dt\right]^{1/p} \leq \left[\int_I k(t)^p dt\right]^{1/p} + 1$$

on the compact interval $I$. From the Minkowski inequality we get

$$\left[\int_I [k(t) + \hat{k}(t)]^p dt\right]^{1/p} \leq 2\left[\int_I k(t)^p dt\right]^{1/p} + 1 \tag{12}$$

for $0 < h \leq h_0$.

Now let $\varepsilon > 0$ be given. We shall show that there is a constant $L > 0$ (depending only on $g$) and on $h_1$, $0 < h_1 \leq h_0$, such that for $0 < h \leq h_1$ one has

$$\int_I |F(x(t), t, h)|^p dt \leq L\varepsilon^p \tag{13}$$

for all $x(\cdot)$ in $\mathcal{K}$.

Since $K$ is compact we can find an $\varepsilon$-net $S = \{x_1, \ldots, x_r\}$ for $K$. That is, for each $x$ in $K$ there is an $x_i$ in $S$ such that $|x - x_i| \leq \varepsilon$. Since (11) holds for $x(t) = x_i$, $i = 1, \ldots, r$, we choose $h_1$ so that $0 < h_1 \leq h_0$ and for $0 < h \leq h_1$ one has

$$\int_I |F(x_i, t, h)|^p dt \leq \varepsilon^p / r, \qquad (i = 1, \ldots, r). \tag{14}$$

Now let $\hat{x}(t)$ be any measurable function defined $R$ with range in $S$. This can be written in the form

$$\hat{x}(t) = \sum_{i=1}^{r} \lambda_i(t) x_i$$

where $\lambda_i(t)$ is 0 or 1 and $\Sigma_{i=1}^r \lambda_i(t) = 1$. For $0 < h \leq h_1$, (14) implies that

$$(15)\qquad \int_I |F(\hat{x}(t), t, h)|^p dt = \int_I \sum_{i=1}^{r} \lambda_i(t)\, |F(x_i, t, h)|^p dt \le \varepsilon^p .$$

Now let $x(\cdot)$ be any continuous function in $\mathcal{K} = C(J, K)$. Then there is a measurable function $\hat{x}(\cdot)$ with range in $S$ such that

$$|x(t) - \hat{x}(t)| \le \varepsilon \qquad (t \in J).$$

One then has

$$|F(x(t), t, h)| \le [k(t) + \hat{k}(t)]\, |x(t) - \hat{x}(t)| + |F(\hat{x}(t), t, h)| .$$

By integrating the last inequality and by using the Minkowski inequality (cf. [9]) together with (12) and (15) one gets

$$\left\{ \int_I |F(x(t), t, h)|^p dt \right\}^{1/p} \le \left\{ 2\left[ \int_I k(t)^p dt \right]^{1/p} + 2 \right\} \varepsilon$$

for $0 < h \le h_1$. This then establishes (13).

Since the topology $\mathcal{T}_b$ is stronger than the topology $\mathcal{T}_c$, the following is a corollary of Theorem III.17.

III.18 COROLLARY. *The mapping* $\pi(g, t) = g_t$ *defines a continuous flow on* $(\mathcal{L}_p, \mathcal{T}_c)$.

One can, in fact, prove the corollary for a class of functions larger than $\mathcal{L}_p(W \times R, R^n)$. Specifically one can drop the Lipschitz condition used in defining $\mathcal{L}_p(W \times R, R^n)$ and replace it with the weaker condition:

For almost all $t$, the function $g(x, t)$ is continuous in $x$.

We shall leave the proof of this assertion as an exercise.

For a more detailed discussion of this flow on $\mathcal{L}_p$, we refer the reader to [20]. For a general discussion of these topologies

$\mathcal{T}_c$ and $\mathcal{T}_b$ we refer the reader to [20] and [30]. The question of compact motions in $\mathcal{L}_p$ can be settled in a manner similar to Theorem III.12.

It is clear that the mapping $(f, t) \to f_t$ $(t \geq 0)$ defines a semiflow on $\mathcal{L}_p(W \times R^+, R^n)$. This will be used in Chapter XI.

III.O. **Two-parameter Functions.** A variation of the translation mapping defines a flow on a function space depending on two parameters. We shall use this flow in the study of Volterra integral equations in Chapter XI. The functions $a(t, s)$ defined below will appear as kernels in these integral equations.

For $1 \leq p < \infty$ we define $\mathcal{Q}_p^*$ as the collection of all measurable functions $a(t, s)$ defined on $R \times R$ with range in $M^n$, the family of $n \times n$ matrices, satisfying the following conditions:

(i) for each compact interval $I \subset R$ and each $t$ in $R$ the mapping

$$x(\cdot) \to \int_I a(t, s)\, x(s) ds$$

of $L_p(I, R^n)$ into $R^n$, is bounded and linear, and

(ii) the mapping $t \to a(t, \cdot)$ is continuous in the weak*-topology on $L_p^*(I, R^n)^n$.

Statement (i) can be reformulated as follows: For every $t$ in $R$ and every compact interval $I \subset R$, each row $a(t, \cdot)$ is an element of $L_p^*(I, R^n)$, the dual of $L_p(I, R^n)$. In this way, $a(t, \cdot)$ itself can be viewed as an element of the direct sum

$$L_p^*(I, R^n)^n = L_p^*(I, R^n) \oplus \cdots \oplus L_p^*(I, R^n)$$

for every compact interval $I \subset R$. The weak*-topology on $L_p^*(I, R^n)^n$ (in Statement (iii)) is induced by the weak*-topology on each component. Note that $L_p^*(I, R^n) = L_q(I, R^n)$ where $q$ satisfies $p^{-1} + q^{-1} = 1$.

A flow on $\mathfrak{A}_p^*$ will now be defined by the mapping

$$\pi(a, \tau) = a_\tau \tag{16}$$

where $a_\tau(t, s) = a(\tau + t, \tau + s)$. The identity property and the group property are evident. Continuity depends, of course, on the topology on $\mathfrak{A}_p^*$.

We define a topology $\mathcal{T}_\omega$ on $\mathfrak{A}_p^*$ as follows: A (generalized) sequence $\{a_n\}$ in $\mathfrak{A}_p^*$ is said to converge to a limit $a$ in $\mathfrak{A}_p^*$ if for every compact interval $I \subset R$ and every $x(\cdot)$ in $L_p(I, R^n)$ one has

$$\left[\int_I [a_n(t, s) - a(t, s)]\, x(s)ds\right] \to 0$$

uniformly for $t$ in compact sets. We leave it as an exercise to show that this topology is uniform. Also the next theorem is left as an exercise. A proof is given in [20].

III.19 THEOREM. *The mapping* $\pi(a, \tau) = a_\tau$ *given by* (15) *defines a continuous flow on* $\mathfrak{A}_p^*$.

The fixed points and periodic points in this flow are interesting. The following assertion is easy to check.

III.20 PROPOSITION (A). *A point* $a$ *in* $\mathfrak{A}_p^*$ *is a fixed point for the flow* $\pi$ *given by* (16) *if and only if*

$$a(t, s) = \alpha(t - s)$$

*for some measurable function* $\alpha(\cdot)$.

(B) *A point* $a$ *in* $\mathfrak{A}_p^*$ *is a periodic point for the flow* $\pi$ *given* by (16) *if and only if* $b(r, s) = a(s+r, s)$ *is periodic in* $s$.

What are the compact motions in this flow? The answer is given in [20]. We will omit a formal statement.

In closing, let us note that the mapping $\pi$ given by (16) defines a continuous flow in another topology $\mathcal{T}_s$. But for this it is necessary to restrict the class $\mathcal{A}_p^*$ somewhat. More precisely, the defining condition (ii) is replaced by

(iii) the mapping $t \to a(t, \cdot)$ is continuous in the norm topology on $L_p^*(I, R^n)^n = L_q(I, R^n)^n$.

The norm topology on $L_q(I, R^n)^n$ is generated by the norm topology on each component. Call this new class $\mathcal{A}_p$. Then define a topology on $\mathcal{A}_p$ by saying that $a_n \to a$ if for every compact interval $I \subset R$ one has

$$\int_I |a_n(t, s) - a(t, s)|^q ds \to 0 , \qquad 1 < q < \infty$$

$$\text{ess sup}\{|a_n(t, s) - a(t, s)| : s \in I\} \to 0, \qquad q = \infty$$

uniformly for $t$ in compact sets. (The number $q$ is defined by $p^{-1} + q^{-1} = 1$.)

In Chapter XI we shall be interested in matrix valued functions $a(t, s)$ that are defined for $0 \leq s \leq t < \infty$. We shall assume that these functions satisfy conditions (i) and (ii), or (i) and (iii). In this case the mapping $(a, \tau) \to a_\tau$ $(\tau \geq 0)$ defines a semi-flow.

*Exercises.* 1. In reference to Example E, show that the metric space $(C, \rho)$ is complete.

2. Show that there exists a motion $f_t$ in the Bebutov dynamical system which is not compact and $\Omega_f = \{0\}$.

3. Another metric on $C$ (Example E) is given by

$$\sigma(f, g) = \sup_{T>0} \{\min(1/T, \sup_{|t| \leq T} |f(t) - g(t)|\} .$$

Show that $\sigma$ is equivalent to $\rho$, that is show that the topology generated by $\sigma$ agrees with that generated by $\rho$.

4. Another metric on $\mathcal{L}_p(R, R^n)$ (Example $L$) is generated as follows: First define $\hat{\sigma}_n$ by

$$\hat{\sigma}_n(f, g) = \sup\left\{ \left( \int_\theta^{\theta+1} |f(\xi) - g(\xi)|^p d\xi \right)^{1/p} : -n \leq \theta \leq n \right\},$$

and then define $\hat{\rho}_n$ and $\hat{\rho}$ as was done in Example E. Show that this metric $\hat{\rho}$ is equivalent to the metric $\rho$ defined in Example L.

5. Show that the metric space $(\mathcal{L}_p(R, R^n), \rho)$ is complete.

6. Show that if $f \in \mathcal{L}_p(R, R^n)$ –Example L– then $f \in \mathcal{L}_r(R, R^n)$ where $1 \leq r \leq p$. Let $H_p(f)$ denote the hull of the flow $f_t$ in $\mathcal{L}_p(R, R^n)$ for $1 \leq p < \infty$. Show that $H_s(f) \subset H_r(f)$ whenever $1 \leq r \leq s < \infty$. Show that if $1 \leq r \leq s < \infty$ and $f_t$ is a compact motion in $\mathcal{L}_s(R, R^n)$ then it is compact in $\mathcal{L}_r(R, R^n)$ and $H_s(f) = H_r(f)$.

7. Let $f(\theta) = g(\theta) + h(\theta)$ for $\theta \in R$. Assume that $g$ is bounded and uniformly continuous and $h \in \mathcal{L}_p(R, R^n)$ where

$$\int_{-\infty}^{\infty} |h(\theta)|^p d\theta < \infty .$$

Show that the motion $f_t$ is compact in $\mathcal{L}_p(R, R^n)$. Assume now that $g$ is bounded and uniformly continuous on $R^+$ and that

$$\int_0^\infty |h(\theta)|^p d\theta < \infty ,$$

and show that the motion $f_t$ is positively compact in $\mathfrak{L}_p(R, R^n)$.

8. Show that the topologies $\mathcal{T}_c$ and $\mathcal{T}_b$ in Example N are uniform topologies. Show that $\mathcal{T}_b$ is metrizable.

9. Another topology on $\mathfrak{L}_p(W \times R, R^n)$ —Example N—is defined by saying $g_n \to g$ if for every compact set $K \subset W$ one has

$$\sup_{x \in K} \int_I |g_n(x, t) - g(x, t)|^p dt \to 0$$

for every compact interval $I$.

Show that the translation mapping $\pi(g, t) = g_t$ defines a continuous flow on $\mathfrak{L}_p$ with this topology.

10. Another topology on $\mathfrak{L}_p(W \times R, R^n)$ is defined by saying that $g_n \to g$ if for every compact set $K \subset W$, every compact set $I \subset R$ and every compact family of functions $\mathcal{K} \subset C(I, W)$ there is a generalized sequence $\{\varepsilon_n\}$ of positive numbers with $\varepsilon_n \to 0$ and

$$\sup_{x(\cdot) \in \mathcal{K}} \int_I |g_n(x(s), s) - g(x(s), s)|^p ds \leq \varepsilon_n ,$$

$$\sup_{x(\cdot) \in C(I, K)} \left\{ \int_I |g_n(x(s), s) - g(x(s), s)|^p ds \right\}^{1/p} \leq m(I) + \varepsilon_n$$

where $m(I)$ is the measure of $I$. Show that $\pi(g, t) = g_t$ defines a continuous flow on $\mathfrak{L}_p$ with this topology.

CHAPTER IV

# NONAUTONOMOUS DIFFERENTIAL EQUATIONS

In this chapter and the next we shall consider two other examples of flows that arise in the study of nonautonomous differential equations and Volterra integral equations. This is a refinement of [28] and [29].

IV.A. **The Basic Construction**. Let $W$ be an open set in $R^n$ and $f \epsilon C(W \times R, R^n)$. We shall say that the differential equation

(1) $$x' = f(x, t)$$

is **admissible** if the solutions of (1) are unique. This means that for every point $(x_0, t_0)$ in $W \times R$ there is precisely one solution $\phi$ of (1) that satisfies $\phi(t_0) = x_0$. Let $\phi(x, f, t)$ denote the solution of (1) that satisfies $\phi(x, f, 0) = x$. Let $I_{(x,f)}$ denote the maximal interval of definition of the solution $\phi(x, f, t)$. Let $\gamma(f)$ denote the trajectory of $f_\tau$ in $C(W \times R, R^n)$, cf. Example III.I. We shall now define a flow on the product space $W \times \gamma(f)$. More precisely, if we let

$$D = \{(x, g; t) \epsilon W \times \gamma(f) \times R: t \epsilon I_{(x,g)}\},$$

then we shall show that a flow $\pi: D \to W \times \gamma(f)$ is given by

(2) $$\pi(x, g; t) = (\phi(x, g, t), g_t) .$$

IV.1. THEOREM. *If* $f \in C(W \times R, R^n)$ *is admissible, then the mapping* $\pi$ *given by* (2) *defines a continuous flow on* $W \times \gamma(f)$.

*Proof:* The identity property is trivial to check. In order to check the group property we assume that $\tau \in I_{(x,\, f)}$ and $\sigma \in I_{\pi(x,\, f\,;\, \tau)}$ and define

$$\phi(t) = \phi(x, f, t), \quad \psi(t) = \phi(\phi(\tau), f_\tau, t), \quad \xi(t) = \phi(t + \tau)\,.$$

Then $\phi$ is the solution of $x' = f(x, t)$ that satisfies $\phi(0) = x$, and $\psi$ is the solution of

$$x' = f_\tau(x, t) = f(x, \tau + t) \tag{3}$$

that satisfies $\psi(0) = \phi(\tau) = \phi(x, f, \tau)$. However, $\xi$ is also a solution of (3) and $\xi(0) = \phi(\tau)$. Hence by the uniqueness of solutions we have

$$\xi(t) = \phi(t + \tau) = \psi(t)$$

for all $t$ in $I_{\pi(x,\, f;\, \tau)}$. Consequently

$$\pi(\pi(x, f;\, \tau);\, \sigma) = \pi(\phi(\tau), f_\tau;\, \sigma) = (\psi(\sigma), f_{\tau+\sigma})$$

$$= (\phi(\tau + \sigma), f_{\tau+\sigma}) = \pi(x, f;\, \tau + \sigma)\,.$$

The continuity of $\pi$ and the maximality of the intervals $I_{(x,\, f)}$ follow from the following lemma of Kamke.

IV.2. LEMMA (Kamke). *Let* $\mathfrak{A}$ *denote the colledtion of all admissible functions* $f$ *in* $C(W \times R, R^n)$. *Then the solution function* $\phi(x, f, t)$ *is continuous on the subset of* $W \times \mathfrak{A} \times R$ *for which it is defined. That is, if* $\{f_n\}$ *is a sequence of admissible functions in* $C(W \times R, R^n)$ *with limit* $f$ *in* $C(W \times R, R^n)$, *where* $f$ *is admissible, if* $\{x_n\}$ *is a sequence in* $W$ *with limit* $x$, *and if* $\{\tau_n\}$ *is a sequence in* $I_{(x_n, f_n)}$ with limit $\tau$, *then* $\tau \in I_{(x,f)}$ *and*

$$\phi(x_n, f_n, \tau_n) \to \phi(x, f, \tau) \ .$$

The proof of this lemma is discussed in Appendix A. The remainder of the proof of Theorem IV.1 is straightforward. QED.

Observe that if (1) is autonomous, that is the function $f$ is independent of $t$, then this flow $\pi$ given by (3) reduces to the flow discussed in Example III.A.

IV.B. **The Extension Problem.** The flow constructed above is still lacking in some structure. The reason for this is that the space $\gamma(f)$ is, in general, not closed. One may then ask whether the flow $\pi$ on $W \times \gamma(f)$ can be extended to a flow on $W \times H(f)$ where $H(f)$ is the hull of $f$ in $C(W \times R, R^n)$. Indeed it seems natural that such an extension be given by

$$\pi(x, g; t) = (\phi(x, g, t), g_t) \tag{4}$$

when $x \in W$, $g \in H(f)$ and $t \in I_{(x,g)}$. Equation (4) does define a flow on $W \times H(f)$ **provided** every $g$ in $H(f)$ is admissible, that is provided that for every $g$ in $H(f)$ the equation $x' = g(x, t)$ has unique solutions.

We shall say that a function $f$ in $C(W \times R, R^n)$ is **regular** or that $f$ satisfies the **strong uniqueness condition**, provided the solutions of $x' = g(x, t)$ are unique for every $g$ in $H(f)$. The next statement is now a consequence of Kamke's Lemma.

IV.3. THEOREM. *Let $f$ be a regular function in $C(W \times R, R^n)$. Then the mapping $\pi$ given by* (4) *defines a continuous flow on $W \times H(f)$.*

The uniqueness conditions we impose seem to be necessary for our view point. We want the flow $\pi$ to be continuous, and this implies that the solutions $\phi(x, g, t)$ be continuous, which in turn implies that the solutions of $x' = g(x, t)$ be unique. Attempts have been made to formulate a topological-dynamical theory for abstract flows that lack uniqueness. See, for example, [6], [24], [26], and [32]. This theory, while greatly advanced in some areas, seems to still be lacking in structure for a systematic exploitation of the properties of solutions of differential equations.

IV.C. **Compact Solutions.** Let $f$ be a regular function in $C(W \times R, R^n)$. For $(x, g) \in W \times H(f)$ let $I_{(x,g)} = (\alpha, \beta)$ be the maximal interval of definition of the solution $\phi(x, g, t)$. We shall say that $\phi(x, g, t)$ is **positively compact** if the set

$$\{\phi(x, g, t) : 0 \leq t < \beta\}$$

lies in a compact set in $W$. **Negative compactness and compactness** are defined similarly. Since $I_{(x,g)}$ is maximal it follows that $\beta = +\infty$ whenever $\phi(x, g, t)$ is positively compact. If this solution is compact, then $I_{(x,g)} = R$. The next lemma, which is easily verified, gives a basic relationship between the solution $\phi(x, g, t)$ and the corresponding motion $\pi(x, g, t)$.

IV.4. LEMMA. *Let $f$ be a regular function in $C(W \times R, R^n)$.*

(A) *If a motion $\pi(x, g, t)$ in $W \times H(f)$ is positively compact* {*compact*}, *then the motion $g_t$ in $H(f)$ is positively compact*

*{compact} and the corresponding solution $\phi(x, g, t)$ is positively compact {compact}.*

(B) *Assume that the motion $f_t$ in $H(f)$ is positively compact {compact}. Then a solution $\phi(x, f, t)$ is positively compact {compact} if and only if the corresponding motion $\pi(x, f, t)$ is positively compact {compact}.*

The natural projections in $W \times H(f)$ given by

$$P: W \times H(f) \to W \text{ and } Q: W \times H(f) \to H(f),$$

where $P(x, g) = x$ and $Q(x, g) = g$, are continuous. Also $Q$ maps the flow $\pi$ given by (4) onto the flow $g_t$ in $H(f)$. The projection $P$ is of great interest when one wants to compare the solutions $\phi(x, g, t)$ and the motions $\pi(x, g, t)$. For example, we note that the motion $\pi(x, g, t)$ is defined for all $t \geq 0$ if and only if the solution $\phi(x, g, t)$ is defined for all $t \geq 0$. In this case the $\omega$-limit set $\Omega_{(x, g)}$ is defined, although it may be empty. We define the **positive limit set** of the solution $\phi(x, g, t)$ as

$$L^+_{(x, g)} = P(\Omega_{(x, g)}) .$$

If the motion $f_t$ in $H(f)$ is positively compact, then the positive limit set $L^+$ can be characterized as follows:

IV.5. LEMMA. *Let $f$ be a regular function in $C(W \times R, R^n)$ and assume that $f_t$ is positively compact. Then a point $y$ lies in the positive limit set $L^+_{(x, f)}$ of the solution $\phi(x, f, t)$ if and only if there is a sequence $\{\tau_n\}$ in $R$ with*

(5) $$\tau_n \to \infty \quad \text{and} \quad \phi(x, f, \tau_n) \to y .$$

*Proof:* Recall that a point $(y, g)$ lies in $\Omega_{(x,f)}$ if and only if there is a sequence $\{\tau_n\}$ in $R$ with

$$\tau_n \to \infty \text{ and } \pi(x, f; \tau_n) \to (y, g) ,$$

cf. Proposition II.6. This last limit can be rewritten as

$$\phi(x, f, \tau_n) \to y \quad \text{and} \quad f_{\tau_n} \to g . \tag{6}$$

If $y \in L^+_{(x,f)}$, then there is a $g$ in $H(f)$ such that $(y, g) \in \Omega_{(x,f)}$. Hence there is a sequence $\{\tau_n\}$ in $R$ satisfying (5). Conversely, let $y \in W$ and let $\{\tau_n\}$ be a sequence in $R$ that satisfies (5). Since the motion $f_t$ is positively compact, we can find a convergent subsequence of $\{f_{\tau_n}\}$, say that $f_{\tau_n} \to g$. Hence (6) is satisfied and $(y, g) \in \Omega_{(x,f)}$, or $y \in L^+_{(x,f)}$. QED.

IV.6. THEOREM. *Let $f$ be a regular function in $C(W \times R, R^n)$ and assume that $f_t$ is positively compact. Let $\phi(x, f, t)$ be a positively compact solution of $x' = f(x, t)$. Then the $\omega$-limit set $\Omega_{(x,f)}$ in $W \times H(f)$ is nonempty, compact and invariant, and the positive limit set $L^+_{(x,f)}$ is a nonempty, compact set in $W$. If $(y, g) \in \Omega_{(x,f)}$ then the solution $\phi(y, g; t)$ is compact. Also, if $y \in L^+_{(x,f)}$ then there is a $g$ in $H(f)$ such that $\phi(y, g, t) \in L^+_{(x,f)}$ for all $t$ in $R$.*

*Proof:* This follows from Theorem IV.3, Lemma IV.4, Theorem II.8 and the fact that the projection $P: W \times H(f) \to W$ is continuous. QED.

**Historical Remark.** The behavior of the positive limit set $L^+_{(x,f)}$ as portrayed in the last theorem is related to dynamical invariance. It was this observation that prompted Miller [34] to introduce a concept of "quasi-invariance."

IV.D. **Limiting Equations.** Recall that the mapping $(f, t) \to f_t$ defines a continuous flow on $C(W \times R, R^n)$. Let $\Omega_f$ denote the $\omega$-limit set of the motion $f_t$. We define the **set of limiting equations** for

$$x' = f(x, t)$$

to be the set of all differential equations of the form

$$x' = f^*(x, t) \qquad (f^* \in \Omega_f) .$$

The next result is a direct consequence of Theorem II.8.

IV.7. PROPOSITION. *Let $f_t$ be a positively compact motion in $C(W \times R, R^n)$. Then the set of limiting equations for $f$ is a nonempty, compact, connected subset of $H(f)$.*

If one refers back to Theorem IV.6, one is able to derive further information about the limiting solution $\phi(y, g, t)$. This is a solution of $y' = g(y, t)$, which is a limiting equation for $x' = f(x, t)$. This fact follows from the observation that the projection

$$Q: \; W \times H(f) \to H(f)$$

maps $\Omega_{(x, f)}$ onto $\Omega_f$.

A form of Theorem IV.6 is valid even without the assumption of positive compactness.

IV.8. THEOREM. *Let $f$ be a regular function in $C(W \times R, R^n)$ and assume that the solution $\phi(x, f, t)$ is defined for all $t \geq 0$. If $(y, g) \in \Omega_{(x, f)}$, then there is a sequence $\{\tau_n\}$ with $\tau_n \to \infty$ and $\phi(x, f, \tau_n + t) \to \phi(y, g, t)$ uniformly for $t$ in compact subsets of $I_{(y, g)}$, the maximal interval of definition of $\phi(y, g, t)$.*

*Proof:* This follows from Proposition II.4, Proposition II.6 and Theorem IV.3. QED.

As we shall see, the concept of the limiting equations will play an essential role in our theory. We shall show how certain properties of the given equation are inherited by the limiting equations and vice versa. This analysis requires a better understanding of the structure of $\omega$-limit sets. Before we can study this we have to introduce such concepts as minimal sets, recurrent motions and almost periodic motions.

IV.E. **Discontinuous Differential Equations.** Consider now a differential equation $x' = f(x, t)$, where $f \in L_p(W \times R, R^n)$, $1 \leq p < \infty$, cf. Example III.N. That is, $f$ is measurable and for every compact set $K \subset W$ there are measurable functions $m(t)$ and $k(t)$ satisfying

$$|f(x, t)| \leq m(t) \qquad (x \in K, \ t \in R)$$

$$|f(x, t) - f(y, t)| \leq k(t)|x - y| \qquad (x, y \in K, \ t \in R)$$

where $m$ and $k$ are locally $L_p$.

The Lipschitz condition on $f$ assures us that the solutions of $x' = f(x, t)$ are unique. We shall use the notation $\phi(x, f, t)$ to denote the solution, as before, and $I_{(x,f)}$ will denote the maximal interval of definition of this solution. One might expect that the mapping $\pi$ given by

$$\pi(x, g; t) = (\phi(x, g, t), g_t) \tag{7}$$

defines a flow on $W \times \gamma(f)$. If one studies the proof of Theorem IV.1 carefully, it becomes evident that $\pi$ will indeed be a flow provided one can generalize Kamke's lemma. This, of course, demands that we place a topology on $L_p(W \times R, R^n)$. Two appropriate topologies, $\mathcal{T}_c$ and $\mathcal{T}_b$, are discussed in Example III.N. One can then prove the following result:

IV.9. LEMMA. *Let* $\{x_n\}$, $\{f_n\}$ *and* $\{\tau_n\}$ *be sequences in* $W$, $L_p(W \times R, R^n)$ *and* $I_{(x_n, f_n)}$ *with* $x_n \to x$ *in* $W$, $\tau_n \to \tau$ *in* $R$ *and* $f_n \to f$ *in* $(L_p, \mathcal{T}_c)$, *or* $(L_p, \mathcal{T}_b)$. *Then* $\tau \in I_{(x,f)}$ *and*

$$\phi(x_n, f_n, t_n) \to \phi(x, f, t) \quad .$$

The proof of this can be found in [19]. Also see Appendix A. With this fact one can then show that $\pi$ is a flow.

IV.10. THEOREM. *Let* $\pi$ *be given by* (7). *Then* $\pi$ *is a flow on* $W \times L_p(W \times R, R^n)$, *where* $L_p(W \times R, R^n)$ *has either topology* $\mathcal{T}_c$ *or* $\mathcal{T}_b$.

Since the Lipschitz condition is built into the definition of $L_p(W \times R, R^n)$, one does not have to worry about the question of regularity that arises in the extension problem in Section IV.B. If one dropped the Lipschitz condition and merely assumed $f$ to be continuous in $x$, theregy enlarging $L_p(W \times R, R^n)$, one could ask whether $\pi$ is a flow on this larger space. One would, of course, restrict one's attention to admissible differential equations only. In this case, one can show that $\pi$ is a flow in the topology $\mathcal{T}_c$, but not (necessarily) in $\mathcal{T}_b$. The reason for this seeming discrepancy is that the proof of Lemma III.17 breaks down when the Lipschitz condition is removed. However, Lemma III.17 is still valid in the topology $\mathcal{T}_c$.

Properties of the flow $\pi(x, f, t)$ that are discussed in statements IV.4 through IV.8 are also true for this particular flow. As a matter of fact, these properties are essentially reformulations of certain topological-dynamical properties in the setting of this flow. In a sense, these properties are somewhat independent of the fact that $f$ belongs to $L_p$ or that $L_p$ has a particular topology as we shall now see.

IV.F. **Abstract Formulation for** $\pi(x, f, t)$. Let $\mathfrak{F}$ denote a space of measurable functions $f(x, t)$ defined on $W \times R$ with range in $R^n$, $W$ a set in $R^n$. We shall say that the function space $\mathfrak{F}$ is **admissible** if

(i) for every $f \in \mathfrak{F}$, the differential equation $x' = f(x, t)$ admits unique solutions, and

(ii) for every $t \in R$, the translate $f_t$ lies in $\mathfrak{F}$.

Let $\phi(x, f, t)$ denote the solution with $\phi(x, f, 0) = x$ and let $I_{(x, f)}$ denote the maximal interval of definition of this solution.

Let $\mathfrak{T}$ be a uniform topology on $\mathfrak{F}$. We shall say that $\mathfrak{T}$ is **admissible** if

(i) the mapping $(f, t) \to f_t$ of $\mathfrak{F} \times R$ into $\mathfrak{F}$ is continuous, and

(ii) the solutions $\phi(x, f, t)$ depend continuously on x, $f$ and $t$ in the following sense: if $x_n \to x$ in $W$, $f_n \to f$ in $\mathfrak{F}$, $\{\tau_n\} \subset I_{(x_n, f_n)}$ and $\tau_n \to \tau$, then $\tau \in I_{(x, f)}$ and

$$\phi(x_n, f_n, \tau_n) \to \phi(x, f, \tau).$$

IV.11. THEOREM. *Let* $\mathfrak{F}$ *be an admissible function class and* $\mathfrak{T}$ *an admissible topology on* $\mathfrak{F}$. *Let*

$$\pi(x, f, t) = (\phi(x, f, t), f_t)$$

*be defined for all* $x \in W$, $f \in \mathfrak{F}$ *and* $t \in I_{(x, f)}$. *Then* $\pi$ *is a flow on* $W \times \mathfrak{F}$.

The proof of this is left as an exercise.

The virtue of this abstract formulation is that it isolates precisely those properties of $\mathfrak{F}$ and $\mathfrak{T}$ that are essential for our purposes. For example, the fact that the topology of uniform convergence on compact sets is admissible on $\mathfrak{A} \cap C(W \times R, R^n)$ (i.e., the admissible differential equations in $C(W \times R, R^n)$) is a direct consequence of Theorem III.6 and Lemma IV.2.

*Exercises.* 1. Show that there exists an admissible function $f$ in $C(W \times R, R^n)$ that does not satisfy the strong uniqueness condition.

2. A differential equation $x' = f(x, t)$ is said to satisfy the global existence property if every solution can be continued for all $t$ in $R$. Show that there is an $f$ in $C(W \times R, R^n)$ with the global existence property but where some $f^*$ in $H(f)$ fails to have the global existence property.

3. Let $f(x, t)$ be a function from $C(W \times R, R^n)$ that satisfies a local Lipschitz condition in $x$, where the Lipschitz constant is independent of $t$. That is, for every compact set $K \subset W$, there is a constant $k$ such that

$$|f(x, t) - f(y, t)| \leq k|x - y| , \qquad (x, y \in K, \ t \in R).$$

Show that $f$ is regular.

4. Let $f$ be a regular function in $C(W \times R, R^n)$. A set $A \subset W$ is said to be **quasi-invariant** (cf. [34]) if for every $y$ in $A$ there is a $g$ in $H(f)$ such that $\phi(y, g, t) \in A$ for all $t$ in $I_{(y,g)}$. Assume that $f$ is bounded and uniformly continuous on sets of the form $K \times R^+$, where $K$ is compact in $W$, and that $\phi(x, f, t)$ is a positively compact solution of $x' = f(x, t)$. Show that the positive limit set $L^+_{(x, f)}$ is quasi-invariant.

5. Let $\mathfrak{F}$ denote the space of all functions $f(x, t)$ in $C(W \times R, R^n)$ that are $C^1$ in $x$. Show that $\mathfrak{F}$ is an admissible class of functions. Define a topology $\mathfrak{T}$ by: $f_n \to f$ in $\mathfrak{F}$ if and only if the functions $\{f_n, D_{x_1} f_n, \ldots, D_{x_n} f_n\}$ converge to $\{f, D_{x_1} f, \ldots, D_{x_n} f\}$ uniformly on compact sets in $W \times R$. ($D_{x_i} f = \partial f / \partial x_i$). Show

that this topology $\mathfrak{T}$ is admissible on $\mathfrak{F}$.

6. Prove Theorem IV.11.

7. Show that part of Lemma IV.5 fails if one does not assume that $f_t$ is positively compact.

# CHAPTER V

## *ALMOST PERIODIC FUNCTIONS*

The purpose of this chapter is to discuss some of the basic properties of almost periodic functions with range in a uniform space. Let $(X, A, \geq, V)$ denote the uniform space. A special case will be of paramount importance. This occurs when $X$ is a linear space and the uniform neighborhood system defines a uniformly linear structure on $X$. This means, that the neighborhoods $V_a$ satisfy: $V_a(x) = x + V_a(0)$. The numbers (i) through (ix) will be used to refer to the nine properties of uniform neighborhood systems defined in Chapter I.

V.A. **Definition of Almost Periodicity.** A subset $E$ of the group $T$ is said to be **relatively dense** if there is a number $\ell > 0$ such that every interval of $T$ of length $\ell$ contains at least one point of $E$. Any such number $\ell$ is called an **inclusion interval** of $E$.

Some examples of relatively dense sets in $R$ are

$$\{n\tau:\ n \in I,\ \ \tau > 0\}$$

and

$$\{\pm\sqrt{n}:\ n \in I^{+}\}\ .$$

Sets which are not relatively dense include $\{\pm n^2:\ n \in I\}$ and $\{\pm p:\ p \text{ is a prime}\}$.

Let $f\colon T \to X$ be a continuous function. For any index $a \in A$ define

(1) $$E(a, f) = \{\tau \epsilon T\colon\ f(t) \epsilon V_a(f(t+\tau)) \text{ for all } t \text{ in } T\}\ .$$

Because of the symmetry property (iii), one also has

$$E(a, f) = \{\tau \epsilon T\colon\ f(t+\tau) \epsilon V_a(f(t)) \text{ for all } t \text{ in } T\}\ .$$

We shall say that $f$ is **almost periodic** if it is continuous and for every $a \epsilon A$ the set $E(a, f)$ is relatively dense in $T$.

If $f\colon\ T \to X$ is continuous and **periodic**, that is if $f(\tau + t) = f(t)$ for all $t$ in $T$ and some $\tau > 0$, then it is clear that $f$ is almost periodic.

If $X$ is a normed linear space and the correspondence $V$ is generated by norm as follows:

$$V_a(x) = \{y \epsilon X\colon\ \|x-y\| < a^{-1}\}\ .$$

where the directed set $(A, \geq)$ is the positive reals with the usual order, then (1) can be rewritten as

(2) $$E(a, f) = \{\tau \epsilon T\colon\ \|f(t+\tau) - f(t)\| < a^{-1} \text{ for all } t \text{ in } T\}\ .$$

In this case the concept of almost periodicity agrees with the concept of **Bohr almost periodicity**, cf. [10].

We shall show below that in a uniform linear space the sum of two almost periodic functions is almost periodic. It follows in such a space, the sum of two periodic functions is almost periodic.

V.B. **Properties of Almost Periodic Functions.** Here we collect some of the basic properties of almost periodic functions which will be needed for the study of minimal sets in the next chapter.

V.1. THEOREM. *Let $\{f_n\}$ be a (generalized) sequence of almost periodic functions defined on $T$ with range in $X$. Assume* *that $f_n$ converges uniformly to a limit function f. Then f is almost periodic.*

*Proof:* Let $a \in A$. We will show that there is a $b \in A$ and an index $N$ so that

$$E(a, f) \supset E(b, f_N) \quad , \tag{3}$$

which would prove that $E(a, f)$ is relatively dense. The continuity of the limit function $f$ follows from the standard arguments of advanced calculus. From this we conclude that $f$ is almost periodic.

First we choose $b \in A$, by Proposition I.4, so that

$$x_i \in V_b(x_{i+1}), \qquad i = 1, 2, 3 \Longrightarrow x_1 \in V_a(x_4) \quad . \tag{4}$$

Because of the uniform convergence of $\{f_n\}$ we can find an index $N$ so that

$$f_n(t) \in V_b(f(t)). \qquad (t \in T) \, , \tag{5}$$

for all $n \geq N$.

If $\tau \in E(b, f_N)$, then one has $f_N(t + \tau) \in V_b(f_N(t))$ for all $t$ in $T$. By applying (4) and (5), together with the symmetry of $V_a$ [property (iii)] we get $f(t) \in V_a(f(t + \tau))$ for all $t$ in $T$. Hence $\tau \in E(a, f)$ which establishes (3). QED.

V.2. THEOREM. *Let X and Y be two uniform spaces and let $g$: $X \to Y$ be uniformly continuous. If $f$: $T \to X$ is almost periodic, then the composition $g \cdot f$: $T \to Y$ is almost periodic.*

*Proof:* Let $A$ be the index set for the uniformity for $X$ and let $A'$ be the index set for the uniformity for $Y$. If $a' \in A'$, choose $a \in A$, by the uniform continuity of $g$, so that

$$g(x') \in V_{a'}(g(x))$$

whenever $x' \in V_a(x)$, for all $x$ in $X$. If $\tau \in E(a, f)$, then

$$f(t) \in V_a(f(t+\tau))$$

for all $t$ in $T$, which implies that $g(f(t)) \in V_{a'}(g(f(t+\tau)))$ for all $t$ in $T$. Hence $E(a, f) \subset E(a', g \cdot f)$. QED.

V.3. THEOREM. *Let $f: T \to X$ be almost periodic. Then $f$ is uniformly continuous.*

*Proof:* Let $a \in A$ and choose $b \in A$, by Proposition I.4, so that

(6) $$x_i \in V_b(x_{i+1}),\ i = 1,2,3 \Longrightarrow x_1 \in V_a(x_4)\ .$$

Let $\ell > 0$ be an inclusion interval for $E(b, f)$. Since $f$ is continuous on $[0, \ell + 1]$ it is uniformly continuous on this interval. Now choose $\delta > 0$ so that

(7) $$f(t_1) \in V_b(f(t_2))$$

whenever $t_1, t_2 \in [0, \ell + 1]$ and $|t_1 - t_2| < \delta$. With $\delta$ so chosen, let $t$ and $s$ be two points in $T$ with $|t - s| < \delta$. Then there is a $\tau$ in $E(b, f)$ so that both $t + \tau$ and $s + \tau$ are in $[0, \ell + 1]$. One then has

(8) $$f(r+\tau) \in V_b(f(r))$$

for all $r$ in $T$. By combining (6), (7) and (8) with the symmetry of $V$, we get $f(t) \in V_a(f(s))$. QED.

**A characterization of almost periodicity.** We give here a characterization of almost periodicity due to Besicovitch. A function $f: T \to X$ is said to be **normal** if it is continuous and for every sequence $\{\tau_n\}$ in $T$ there is a subsequence $\{\tau_n'\}$ such that the sequence of functions $\{f_{\tau_n'}\}$ is a **uniform Cauchy sequence.** This means that for every $a \in A$ there is an index $N$ so that

$$f(t+\tau_n') \in V_a(f(t+\tau_m'))$$

for $t \in T$ and $n,m \geq N$.

In a complete uniform space the concept of normality can be reformulated as follows:

V.4. PROPOSITION. *Let $X$ be a complete uniform space. A continuous function $f\colon T \to X$ is normal if and only if for every sequence $\{\tau_n\}$ in G, there is a subsequence $\{\tau_n'\}$ such that the sequence of functions $\{f_{\tau_n'}\}$ is uniformly convergent.*

This proposition asserts that there is a function $g\colon T \to X$ such that $f_{\tau_n'}(t) \to g(t)$ uniformly for $t$ in $T$. It follows that $g$ is continuous.

V.5. THEOREM. (Besicovitch) *Let $f\colon T \to X$. The function $f$ is normal if and only if it is almost periodic.*

*Proof:* First we assume that $f$ is almost periodic. Let $\{\tau_n\}$ be a sequence in $T$ and $a \in A$. Choose $b \in A$, by Proposition I.4, so that

$$x_i \in V_b(x_{i+1}),\ i = 1,2,3,4 \implies x_1 \in V_a(x_5) \ . \tag{9}$$

Let $\ell > 0$ be an inclusion interval for the relatively dense set $E(b,f)$. Then any $\tau_n$ can be written as

$$\tau_n = \rho_n + \sigma_n$$

where $\sigma_n \in E(b,f)$ and $0 \leq \rho_n \leq \ell$. (This representation for $\tau_n$

is not unique, so we choose one such representation for every $\tau_n$.) Now let $\rho$ be an accumulation point of the sequence $\{\rho_n\}$. By Theorem V.3, $f$ is uniformly continuous, so choose $\delta > 0$ so that $f(t) \in V_b(f(s))$ whenever $|t - s| \leq 2\delta$. Since $\rho$ is an accumulation point of $\{\rho_n\}$, the set of all $\tau_n$ with $\rho - \delta \leq \rho_n \leq \rho + \delta$ forms a subsequence, which we denote by $\{\tau_n'\}$. We now claim that

$$f(t + \tau_n') \in V_a(f(t + \tau_m'))$$

for all $t$ in $T$. This is established by applying (9) to

$$x_1 = f(t + \tau_n') ,$$

$$x_2 = f(t + \sigma_n') ,$$

$$x_3 = f(t) ,$$

$$x_4 = f(t + \sigma_m') \quad \text{and}$$

$$x_5 = f(t + \tau_m') .$$

Hence $f$ is normal.

Now assume that $f$ is normal. If $f$ is not almost periodic then there is an $a \in A$ such that $E(a, f)$ is not relatively dense. Now let $\tau_1 > 0$ and let $(a_2, b_2)$ be an interval of length $> 2|\tau_1|$ such that

$$(a_2, b_2) \cap E(a, f) = \emptyset ,$$

and let $\tau_2$ denote the center of this interval. Then $\tau_2 - \tau_1 \in (a_2, b_2)$ and $\tau_2 - \tau_1 \notin E(a, f)$. Now define an interval $(a_3, b_3)$ of length $> 2(|\tau_1| + |\tau_2|)$ with

$$(a_3, b_3) \cap E(a, f) = \emptyset ,$$

and let $\tau_3$ denote the center of this interval. Then $\tau_3 - \tau_1$ and

$\tau_3 - \tau_2$ are not in $E(a, f)$. In a similar way define $\tau_4, \tau_5, \ldots$ so that none of the numbers $\tau_i - \tau_j$ belong to $E(a, f)$. This means that

$$f(t + \tau_i - \tau_j) \notin V_a(f(t)) \qquad (i \neq j)$$

for some $t$, or equivalently

$$f(t + \tau_i) \notin V_a(f(t + \tau_j)) \qquad (i \neq j)$$

for some $t$. But this implies that no subsequence of $\{f_{\tau_i}\}$ is uniformly Cauchy. QED.

V.6. COROLLARY. *Let $f\colon T \to X$ be an almost periodic function. Then the range of $f$, $f(T)$, is a totally bounded set in $X$. In particular if $X$ is complete, then $f(T)$ is contained in a compact set.*

*Proof:* We shall use the fact that a set $A \subset X$ is totally bounded if every (generalized) sequence $\{x_n\}$ in $A$ has a subsequence that is a Cauchy sequence. Let $\{f(\tau_n)\}$ denote an arbitrary sequence in $f(T)$ where $\{\tau_n\} \subset T$. Since $f$ is normal we can find a subsequence $\{\tau_n'\}$ such that the sequence of translates $\{f_{\tau_n'}\}$ is uniformly Cauchy, which implies that $f_{\tau_n'}(0) = f(\tau_n')$ is a Cauchy sequence. QED.

V.7. COROLLARY. *Let $f\colon T \to X$ be an almost periodic function and let $\{\tau_n\}$ be a sequence in $T$. If $f_{\tau_n} \to f^*$ uniformly on compact sets in $T$, then $f^*$ is almost periodic and $f_{\tau_n} \to f^*$ uniformly in $T$.*

*Proof:* There is no loss in generality in assuming that $X$ is complete, since we can always imbed $X$ in its completion. This imbedding does not destroy the almost periodicity of $f$.

Since $f$ is normal, let us choose any subsequence of $\{f_{\tau_n}\}$ that is uniformly convergent, for simplicity call this subsequence $\{f_{\tau_n}\}$. Then there is a continuous function $g$ such that $f_{\tau_n} \to g$ uniformly. Since $f_{\tau_n} \to f^*$ uniformly on compact sets, it follows that $f^* = g$. The result now follows from Theorem V.1 QED.

V.C. **Almost Periodic Functions in Linear Spaces.** What more can be said about almost periodic functions when the range $X$ is a linear space with a uniformly linear structure?

V.8. LEMMA. *Let $f\colon T \to X$ and $g\colon T \to X$ be almost periodic functions where $(X, A, V)$ is a linear space with a uniformly linear structure. Then*

$$E(a, f) \cap E(a, g)$$

*is relatively dense in $T$ for every $a \in A$.*

*Proof:* Let $a \in A$ and choose $b \in A$ by (viii) so that $a \leq b$ and if $x, y \in V_b(0)$ then $x + y \in V_a(0)$. Let $\delta > 0$. We will first show that the set

$$E(a, f) \cap S(E(a, g), \delta) \tag{10}$$

is relatively dense in $T$, where

$$S(E(a, g), \delta) = \{\sigma \in T\colon |\tau - \sigma| \leq \delta \text{ for some } \tau \text{ in } E(a, g)\}.$$

To do this we let $\ell = k\delta$ ($k$ an integer) be an inclusion interval for both $E(b, f)$ and $E(b, g)$. Let $I_n = [(n-1)\ell, n\ell]$, $n = 0, \pm 1, \ldots$ In each $I_n$ there exist numbers $\sigma_n$ and $\tau_n$ belonging to $E(b, f)$ and $E(b, g)$, respectively. Therefore $|\sigma_n - \tau_n| \leq \ell$. Now define

$$J_i = [(i-1)\delta, i\delta), \qquad i = -k+1, \ldots, k-1,$$
$$J_k = [(k-1)\delta, k\delta].$$

Then $\sigma_n - \tau_n$ belongs to precisely one of these intervals $J_i$, call it $J(n)$. Since the number of intervals $J_i$ is finite, there is obviously an integer $n_0 > 0$ with the following property:

For every integer $n$, there corresponds an integer $n'$ such that

$$|n'| \leq n_0 \text{ and } J(n) = J(n') .$$

This means that there is a $\theta$, $|\theta| \leq 1$ with

$$\begin{aligned} &\sigma_n - \tau_n = \sigma_{n'} - \tau_{n'} + \theta\delta , \text{ or} \\ (11) \qquad &\sigma_n - \sigma_{n'} = \tau_n - \tau_{n'} + \theta\delta . \end{aligned}$$

Since $\sigma_n$ and $\sigma_n'$ are in $E(b, f)$ it follows that $\sigma_n - \sigma_n'$ is in $E(a, f)$. To see this we note that $V_a(0) = -V_a(0)$ for all $a$ in $A$ and that

$$f(t) \in V_b(f(t-\sigma_n')), \; f(t-\sigma_n') \in V_b(f(t+\sigma_n - \sigma_n')) \Longrightarrow$$
$$f(t) \in V_a(f(t+\sigma_n - \sigma_n')) .$$

Similarly one has $\tau_n - \tau_n'$ in $E(a, g)$. It follows from (11) that every number of the form $\rho_n = \sigma_n - \sigma_n'$ is also in $S(E(a, g), \delta)$. However,

$$|\rho_n - \rho_{n+1}| \leq (2n_0 + 3)\ell$$

therefore the set in (10) is relatively dense in $T$. We emphasize that this holds for every $a$ in $A$ and every $\delta > 0$.

With $a$ and $b$ (in $A$) given as above we shall now show that the set

$$E(a, f) \cap E(a, g)$$

is nonempty and relatively dense in $T$. Since $g$ is uniformly continuous (Theorem V.3) one can find a $\delta > 0$ such that

$$g(t) \in V_b(g(\tau + t)) \,, \qquad (t \in T),$$

whenever $|\tau| \leq \delta$. We claim that if $\sigma \in E(b, g)$ and $|\tau| \leq \delta$ then $\sigma + \tau \in E(a, g)$. Indeed, one has

$$g(t) \in V_b(g(\tau + t)),\ g(\tau + t) \in V_b(g(\sigma + \tau + t)) \implies$$

$$g(t) \in V_a(g(\tau + \sigma + t)) \ .$$

This means that $S(E(b, g), \delta) \subset E(a, g)$. Since $E(b, f) \subset E(a, f)$ it follows that $E(a, f) \cap E(a, g)$ is relatively dense in $T$. QED.

In the exercises, the reader is asked to prove the analogue of the last lemma for the case where the range of $f$ and $g$ lie in different uniform linear spaces. The next result is then a direct consequence of this more general statement. We note that if $X$ and $Y$ are two uniform linear spaces with uniformities generated by $V_a$ and $V_b$ ($a \in A$ and $b \in B$), respectively, then the product $X \times Y$ is a uniform linear space with uniformity generated by $V_{(a,b)}$ where $V_{(a,b)} = V_a \times V_b$. (The index set for $X \times Y$ is $A \times B$ with the usual product ordering.)

V.9. THEOREM. *Let $f\colon T \to X$ and $g\colon T \to Y$ be continuous and define $h\colon T \to X \times Y$ by $h(t) = (f(t), g(t))$. Assume that $X$ and $Y$ are uniform linear spaces. Then $h$ is almost periodic if and only if both $f$ and $g$ are almost periodic.*

*Proof:* First we note that the natural projections $P\colon X \times Y \to X$ and $Q\colon X \times Y \to Y$ are uniformly continuous, and that $f = P \circ h$ and $g = Q \circ h$. If $h$ is almost periodic, then $f$ and $g$ are almost periodic by Theorem V.2.

For the converse we consider the case where $X = Y$. (The general case is a consequence of Exercise 1, below.) Let $B = A \times A$ denote the index set for $X \times Y$, and choose an index $(a, b)$. Then choose an index c in $A$ with $a \leq c$ and $b \leq c$. It follows then that $(a, b) \leq (c, c)$. It is easily seen that

$$E(c, f) \cap E(c, g) \subset E((c, c), h) \subset E((a, b), h) \quad . \tag{12}$$

If $f$ and $g$ are almost periodic, it follows from (12) and Lemma V.8, that $h$ is almost periodic. QED.

V.10. THEOREM. *Let $f\colon T \to X$ and $g\colon T \to X$ be almost periodic functions, where $X$ is a uniform linear space $X$. Then $f + g$ is almost periodic.*

*Proof:* Let $a \in A$ and choose $b \in A$ so that $a \leq b$ and $x + y \in V_a(0)$ whenever $x, y \in V_b(0)$. It follows then that

$$E(b, f) \cap E(b, g) \subset E(a, f+g) \quad ,$$

and this theorem then follows from Lemma V.8. QED.

V.11. COROLLARY. *Let $X$ be a complex uniform linear space, Then every trigonometric polynomial*

$$f(t) = \sum_{n=1}^{N} e^{i\lambda_n t} a_n \quad ,$$

*where $\{a_1, \ldots, a_N\}$ are in $X$ and $\{\lambda_1, \ldots, \lambda_N\}$ are real, is almost periodic.*

*Proof:* Each term $e^{i\lambda_n t} a_n$ is periodic in $t$ and continuous. Hence it is almost periodic. The corollary now follows from Theorem V.10. QED.

V.12 COROLLARY. *Let $X$ be a complex uniform linear space. Then a uniformly convergent series*

$$f(t) = \sum_{n=1}^{\infty} e^{i\lambda_n t} a_n$$

*where $a_n \in X$ and $\lambda_n$ are real $(n = 1, 2, \ldots)$, is an almost periodic function.*

*Proof:* This follows from Corollary V.11 and Theorem V.1. QED.

The author feels that almost periodic functions are of practical interest because they admit a Fourier series expansion of the type discussed above. One can show that every almost periodic function $f\colon T \to X$, where $X$ is a complete, complex uniform linear space admits a unique Fourier series expansion.

V.D. **An important example.** Let $X = C(W, B)$ denote the space of continuous functions defined on $W$ with range in $B$, where $W$ is an open set in $R^n$ and $B$ is a Banach space. We shall assume that $X$ has the topology of uniform convergence on compact sets. This topology is metrizable, and a metric $\rho$ can be constructed as follows: Let $\{K_n\}$ be a sequence of compact sets in $W$ with $K_n \subset K_{n+1}$ and $W = \cup_n K_n$. If $f, g \in X$ define

$$\sigma_n(f, g) = \sup\{\|f(x) - g(x)\| : x \in K_n\}$$

$$\rho_n(f, g) = \min(1, \sigma_n(h, g)), \qquad \rho(f, g) = \sum_{n=1}^{\infty} 2^{-n}\rho_n(f, g) .$$

The function $\rho$ is a metric on $X$ and, moreover, $X$ is complete. If $\phi\colon T \to X$ is a continuous function, we shall denote the value of $\phi$ at $t$ by $\phi^t$. Then for each $f$ in $X$, $\phi^t(f)\colon W \to B$. We shall define $f(x, t)$ by

$$\phi^t(f) : x \to f(x, t) \tag{13}$$

where $x \in W$. It follows that $f(x, t)$ is jointly continuous in $x$ and $t$, that is $f(x, t) \in C(W \times T, B)$. Equation (13) defines a one-to-one correspondence between continuous mappings $\phi : T \to X$ and the elements of $C(W \times T, B)$. One may ask when is $\phi$ almost periodic. The answer, which we leave as an exercise (cf. Theorem VI.18), is as follows:

The function $\phi : T \to X$ is almost periodic if and only if

(i) the function $f$ is uniformly continuous on sets of the form $K \times T$ where $K$ is compact in $W$, and

(ii) for each $x$ in $X$, the function $f(x, \cdot) : T \to B$ is Bohr almost periodic.

*Exercises* 1. Let $f : T \to X$ and $g : T \to Y$ be almost periodic functions where $X$ and $Y$ are uniform linear spaces. Show that the set $E(a, f) \cap E(b, g)$ is relatively dense in $T$, where $a$ and $b$ are any two indices from the index sets which define the uniform structure on $X$ and $Y$, respectively.

2. Prove Theorem V.9 for the case where $X \neq Y$.

3. Let $X$ be a complex Banach space, with the uniformity generated by the norm $\| \cdot \|$. Show that the collection of all almost periodic functions $f : T \to X$ forms a Banach space with the norm given by

$$\|f\|_\infty = \sup\{\|(f(t)\| : t \in T\} .$$

4. Analyze the example of Section V.D in the case where the Banach space $B$ is replaced by a uniform space.

5. Let $F$ be a subset of the Banach space of almost periodic functions defined in Exercise 3. For $a > 0$ define $E(a, F)$ by

$$E(A, F) = \{\tau \in T: \|f_\tau - f\|_\infty < a^{-1}, f \in F\}$$

(i) Show that if $E(a, F)$ is relatively dense in $T$ for every $a > 0$, then the family $F$ is equicontinuous. (ii) Assume that $F$ is bounded. Show that the closure $\bar{F}$ is compact if and only if $E(a, F)$ is relatively dense for every $a > 0$.

## CHAPTER VI

## *RECURRENT MOTIONS AND ALMOST PERIODIC MOTIONS*

We now return to the study of dynamical systems or flows on a uniform space $X$. One of our objectives here is to prove the Birkhoff Recurrence Theorem, which gives a characterization of compact minimal sets. Three concepts will play a fundamental role in this chapter, viz., Poisson stable motions, recurrent motions and almost periodic motions.

VI.A. **Poisson stability.** Let $X$ be a uniform space and let $\pi$ be a flow on $X$. A motion $\pi(x, t)$ is said to be **positively Poisson stable** (**stable** $P^+$) if it is defined for all $t \geq 0$ and for every neighborhood $U$ of $x$ there is a (generalized) sequence $\{t_n\}$ with $t_n \to +\infty$ and $\pi(x, t_n) \in U$. **Negative Poisson stability** (**stable** $P^-$) is defined by demanding that $t_n \to -\infty$. A motion is **Poisson stable** (**stable** $P$) if it is both positively and negatively Poisson stable. A point $x$ in $X$ is said to be **Poisson stable** if the corresponding motion $\pi(x, t)$ is Poisson stable.

VI.1. LEMMA. *Let $x$ be a point in $X$ that is either stable $P^+$, stable $P^-$, or stable $P$, then every point $y$ in $\gamma(x)$ is, respectively, stable $P^+$, stable $P^-$, or stable $P$.*

*Proof:* Let us consider the case where x is stable $P^+$ and let $y = \pi(x, \tau)$ be a point in $\gamma(x)$. Then $U_\tau = \pi(U, \tau)$ is a neighborhood of $y$ whenever $U$ is a neighborhood of x. Furthermore, $\pi(x, t_n + \tau) \in U_\tau$ whenever $\pi(x, t_n) \in U$. From this we conclude that $y$ is stable. $P^+$. The same argument applies for the other cases. QED.

VI.2. THEOREM (A) *The following statements are equivalent:*

(1) x *is stable* $P^+$.

(2) $x \in \Omega_x$.

(3) $\Omega_x = H(x) = \text{Cl}\, \gamma(x)$.

(B) *The following statements are equivalent:*

(1) x *is stable* $P^-$.

(2) $x \in A_x$.

(3) $A_x = H(x)$.

(C) *The point* x *is stable* $P$ *if and only if* $A_x = \Omega_x = H(x)$.

*Proof:* Statement (C) follows from (A) and (B). We shall prove (A) and note that the proof of (B) requires only an obvious modification.

Assume that x is stable $P^+$. Let $V_a(x)$, $a \in A$, be the family of (open) neighborhoods of x generated by the uniformity on $X$. Then $\{x\} = \cap_a V_a(x)$. For each $a \in A$, let $\{t_{an}\}$ be a generalized sequence with $\pi(x, t_{an}) \in V_a(x)$ and $t_{an} \to \infty$. (The index $n$ lies in some directed set $N_a$, which depends on $a$, and the limit $t_{an} \to \infty$ is taken in reference to the order on $N_a$). Now form the product directed set $D = A \times \Pi_a N_a$ with the usual order, cf., Section I.B. It follows from the theorem on iterated limits (Proposition I.3) that $t_{an} \to \infty$ in reference to the order on $D$. Furthermore, $\pi(x, t_{an}) \to x$. Hence $x \in \Omega_x$.

Now assume that $x \in \Omega_x$. Then $\gamma(x) \subset \Omega_x$ since $\Omega_x$ is an invariant set. Furthermore $H(x) = \mathrm{Cl}\,\gamma(x) \subset \Omega_x$ since $\Omega_x$ is a closed set. On the other hand, it follows from the definition of $\Omega_x$ that $\Omega_x \subset H(x)$. Therefore, $\Omega_x = H(x)$.

Now assume that $\Omega_x = H(x)$. It follows from Proposition II.6 that there is a generalized sequence $\{t_n\}$ with $t_n \to \infty$ and $\pi(x, t_n) \to x$. Therefore, if $U$ is any neighborhood of x, then there is a subsequence $\{t_{n'}\}$ with $\pi(x, t_{n'}) \in U$, which implies that x is stable $P^+$.

It should be clear that every periodic point is stable $P$. The following result illustrates what can be said about points that are stable $P^+$ but not periodic.

VI.3. THEOREM. *Let $X$ be a complete uniform space, and let x be a point in $X$ that is stable $P^+$. If x is not a periodic point then*

$$\mathrm{Cl}(H(x) - \gamma(x)) = H(x) = \Omega_x ,$$

*that is, the set of points in $H(x) - \gamma(x)$ is dense in $\Omega_x$.*

*Proof:* Assume that x is not a periodic point. Since the set $H(x) - \gamma(x)$ is invariant, its closure is invariant. If we can show that

$$x \in \mathrm{Cl}(H(x) - \gamma(x)) \tag{1}$$

it will then follow that

$$\Omega_x = H(x) \subset \mathrm{Cl}(H(x) - \gamma(x)) .$$

However, one always has $\mathrm{Cl}(H(x) - \gamma(x)) \subset H(x)$, so the theorem follows once we prove (1). In order to prove (1), we will show that for every index $a$, there is $y \in H(x) - \gamma(x)$ with $y \in V_a(x)$. The point $y$ will be chosen as the limit of an appropriate Cauchy sequence.

Let $\{t_a\}$ be a sequence (defined on $A$) with range in $T^+$ and such that $t_a \to \infty$ and $\pi(x, t_a) \to x$. Let $a$ be a fixed index in $A$ and choose $d$, by Proposition I.5, so that $\mathrm{Cl}\, V_d(x) \subset V_a(x)$. Then for every $b \in A$ define

$$E_b = \{\tau \in T:\ \tau \geq t_b \text{ and } V_b(\pi(x,\tau)) \subset \mathrm{Cl}\, V_d(x)\}\ .$$

Let $B$ denote those indices $b$ in $A$ for which $E_b$ is unbounded in $T$. Since $V_a(x)$ is an open set and $x$ is stable $P^+$ it follows that $B$ is a directed set. Furthermore, the set $E_b$ is closed for every $b$ in $B$. (To see that $E_b$ is closed we let $\{\tau_n\}$ be chosen in $E_b$ with $\tau_n \to \tau$. Since $\pi(x, \tau_n) \to \pi(x, \tau)$, it follows from Proposition I.7 that $V_b(\pi(x, \tau)) \subset \mathrm{Cl}\, V_d(x)$.)

For each ordered pair of indices $(b, c)$ $(b \in B,\ \ c \in A)$ define

$$F_{(b,c)} = \{\tau \in E_b:\ V_c(\pi(x,\tau)) \cap \pi(x, [-t_b, t_b]) = \emptyset\}\ .$$

It follows from Proposition I.7 that $F_{(b,c)}$ is closed. For $b \in B$, let $C_b$ denote those indices c in $A$ for which $F_{(b,c)}$ is unbounded in $T$. It follows from the fact that $x$ is not a periodic point that $C_b$ is a directed set.

Let $D$ be the subset of the directed product set $A \times A$ consisting of all elements of the form $\alpha = (b, c)$ where $b \in B$ and $c \in C_b$. It is easy to show that $D$ is a directed set itself, in the product ordering.

Consider now the collection of ordered pairs $(H, \tau)$ where $H$ is a subset of $D$ and $\tau$ is a function mapping $H$ into G with the property that $\tau_\alpha \in F_\alpha$ for all $\alpha \in H$. Let $y_\alpha = \pi(x, \tau_\alpha)$ for $\alpha \in H$. We shall say that a pair $(H, \tau)$ has property $O$ if

(2) $\quad y_{\alpha'} \in \mathrm{Cl}\, V_{\alpha}(y_{\alpha}), \quad$ whenever $\alpha, \alpha' \in H$ and $\alpha' \geq \alpha$,

where $V_{\alpha}$ is an open neighborhood satisfying $\mathrm{Cl}\, V_{\alpha} \subset V_b \cap V_c$ and $\alpha = (b, c)$. Let $\mathcal{H}$ denote the collection of all ordered pairs $(H, \tau)$ that have proprety $O$. We note that $\mathcal{H}$ is nonempty. Define an ordering on $\mathcal{H}$ by saying that $(H_1, \tau_1) \leq (H_2, \tau_2)$ whenever $H_1 \subset H_2$ and $\tau_2$ is an extension of $\tau_1$. By Zorn's lemma the collection $\mathcal{H}$ has a maximal element, which we will denote by $(H, \tau)$. We claim that $H$ is a direct subset of $D$. To prove this, one must show that $H$ has the upper bound property. If $H$ does not have the upper bound property, then there are two elements $\alpha_1 = (b_1, c_1)$ and $\alpha_2 = (b_2, c_2)$ in $H$ with the property that any $\alpha$ satisfying $\alpha_1 \leq \alpha$ and $\alpha_2 \leq \alpha$ is not in $H$. Now let $\sigma = \max(\tau_{\alpha_1}, \tau_{\alpha_2})$ and $V = V_{\alpha_1} \cap V_{\alpha_2}$. Since $y = \pi(x, \sigma)$ is stable $P^+$ (Lemma VI.1) there is a $\tau_0$ arbitrarily large such that $\pi(y, \tau_0) \in V(y)$. In particular $\tau_0$ can be chosen to lie in $E_b$ for some $b$. Since $E_b$ increases as $b$ increases, we can assume that $\tau_0 \in E_b$ where $b_1 \leq b$ and $b_2 \leq b$. With $b$ so chosen, we can now choose $\tau_0$ to lie in $F_{(b,c)}$ for some c. Since $F_{(b,c)}$ increases in c as c increase, we can assume that $c_1 \leq c$ and $c_2 \leq c$. Then $\alpha = (b, c)$ satisfies $\alpha_1 \leq \alpha$ and $\alpha_2 \leq \alpha$. Furthermore by defining $\tau_{\alpha} = \tau_0$ we see that the function $\tau$ can be extended to $H \cup \{\alpha\}$. This contradicts the maximality of $(H, \tau)$.

Since $H$ is a directed set satisfying (2) we see that the sequence $\{y_{\alpha}\}$ is a Cauchy sequence. Since $X$ is complete, this sequence converges, say that $y_{\alpha} \to y$. Since $y_{\alpha} \subset \mathrm{Cl}\, V_d(x)$, it follows that $y \in \mathrm{Cl}\, V_d(x) \subset V_a(x)$. It remains only to show that $y \notin \gamma(x)$. Assume on the contrary that $y \in \gamma(x)$, then $y = \pi(x, \tau)$ for some $\tau$. One can then find an $\alpha = (b, c)$ such that $|\tau| < t_b$. Hence $y \in \pi(x, [-t_b, t_b])$. However, it follows from (2) that

$y \in \mathrm{Cl}\, V_\alpha(y_\alpha)$. This contradicts the fact that $\tau_\alpha \in F_\alpha$ and $\mathrm{Cl}\, V_\alpha(y_\alpha) \cap \pi(x, [-t_b, t_b]) = \emptyset$. QED.

VI.4. COROLLARY. *Let $\pi(x, t)$ be a stable $P^+$ motion in a complete uniform space $X$. Then $\pi(x, t)$ is periodic if and only if $H(x) = \gamma(x)$.*

*Proof:* If $\pi(x, t)$ is periodic, then $\gamma(x)$ is the continuous image of a compact set $0 \leq t \leq \tau$, where $\tau$ is the period of $\pi(x, t)$. Hence $\gamma(x)$ is compact and closed, that is $H(x) = \mathrm{Cl}\, \gamma(x) = \gamma(x)$. (This assertion is true even when $X$ is not complete.) The converse statement is merely the contrapositive of Theorem VI.3. QED.

VI.B. **Recurrence.** Let $X$ be a uniform space and let $\pi$ be a flow on $X$. A motion $\pi(x, t)$ is said to be **recurrent** if it is defined for all $t$ and for every neighborhood $U$ of $x$, the set

$$\{\tau \in T : \pi(x, \tau) \in U\}$$

is relatively dense in $T$. A point $x$ in $X$ is said to be **recurrent** if the corresponding motion $\pi(x, t)$ is recurrent.

Using the ideas in Lemma VI.1 one easily can prove the following:

VI.5. LEMMA. *Let $x$ be a recurrent point in $X$. Then every point $y$ in $\gamma(x)$ is recurrent.*

The next result gives a characterization of compact, recurrent motions:

VI.6. THEOREM (Birkhoff Recurrence Theorem). *A set $E$ in $X$ is a compact minimal set if and only if $E = H(x)$ for some compact recurrent point $x$.*

*Proof:* Let $E$ be a compact, minimal set in $X$, $x \in E$, and assume that $\pi(x, t)$ is not recurrent. (The motion $\pi(x, t)$ is obviously compact.) Then there exists a nonempty open neighborhood $U$ of $x$, in the relative topology on $E$, and a sequence of intervals $I_n = [t_n - T_n, t_n + T_n]$ with $T_n \to \infty$ and $\pi(x, I_n) \cap U = \emptyset$. Since $E$ is compact, the sequence $x_n = \pi(x, t_n)$ has a convergent subsequence, say that $x_n \to y$. We now claim that $\gamma(y) \subset E - U$. Indeed, if $t \in T$, then by the continuity of $\pi$ one has $\pi(x_n, t) \to \pi(y, t)$. Now choose a subsequence of $\{T_n\}$, denote it by $\{T_n\}$ again, so that $|t| < T_n$. Then

$$\pi(x_n, t) \in \pi(x, I_n) \subset E - U,$$

so the limit point $\pi(y, t)$ lies in the closed set $E - U$. That is, $\gamma(y) \subset E - U$, and consequently $H(y) \subset E - U$. In other words, $H(y)$ is a proper subset of $E$ that is nonempty, closed and invariant, which contradicts the minimality of $E$.

Now let $\pi(x, t)$ be a compact, recurrent motion. It follows that the hull $H(x)$ is compact. In order to show that $H(x)$ is minimal we first note that a set $E$ is minimal if and only if $E = H(y)$ for every $y$ in $E$. Let $y$ be an arbitrary point in $H(x)$. Since $\gamma(x)$ is dense in $H(x)$, it will suffice to show that

$$x \in H(y) . \tag{3}$$

In order to prove (3) we shall show that for every neighborhood $U$ of $x$, there is a point $s$ in $T$ such that $\pi(y, s) \in U$. Let $U$ be given and choose a neighborhood $V$ of $x$ such that $V \subset \mathrm{Cl}\, V \subset U$ (cf. Proposition I.5). Let $\ell > 0$ be an inclusion interval for the relatively dense set $\{\tau \in T: \pi(x, \tau) \in V\}$. Since $y \in H(x)$, there is a sequence $\{t_n\} \subset T$ such that $x_n = \pi(x, t_n) \to y$. Since $x$ is recurrent, there is an $s_n$, $0 \leq s_n \leq \ell$ such that $\pi(x_n, s_n) \in V$. The sequence $\{s_n\}$ has a convergent subsequence, say that $s_n \to s$. It follows from the continuity of $\pi$ that $\pi(x_n, s_n) \to \pi(y, s)$ and $\pi(y, s) \in \mathrm{Cl}\, V \subset U$. QED.

It should be noted that we have actually shown that every motion in a compact, minimal set is necessarily recurrent. This is an extension of Lemma VI.5.

**Historical comment:** Nemytskii and Stepanov [22], in following the treatment of G. D. Birkhoff, define recurrence for metric spaces as follows: A motion $\pi(x, t)$ is recurrent if for every $\varepsilon > 0$ there is a $\tau > 0$ such that for any $t_0$ in $T$ one has

$$\gamma(x) \subset S(\pi(x, [t_0, t_0 + \tau]), \varepsilon)$$

where $S(A, \varepsilon)$ denotes the $\varepsilon$-sphere about $A$. They then show that the two concepts of recurrence are equivalent in the case where the motion $\pi(x, t)$ is compact. Furthermore, if the metric space is complete, then a motion that is recurrent in the sense of Nemytskii and Stepanov is also compact. As a consequence of these facts and Birkhoff's Theorem, we see that if $\pi(x, t)$ is a recurrent motion in the sense of Nemytskii and Stepanov and if the space $X$ is a complete metric space, then $H(x)$ is a compact, minimal set.

It should also be noted that some authors, see [13] for example, use the term "recurrent" to refer to "Poisson stability" and "point-wise almost periodic" to refer to our concept of recurrence.

VI.7. THEOREM. *Let $E$ be a compact, minimal set in $X$. If $E$ has an interior point, then $E$ is an open set, in particular $E$ is a component of $X$.*

*Proof:* Let x be an interior point of $E$ and let $U$ be an open neighborhood of x that is in $E$. Then

$$V = \bigcup_{t \in T} \pi(U, t)$$

is an open set. Moreover, we claim that $V = E$. Indeed, if $y \in E$, then $\gamma(y)$ is dense in $E$. Hence there is a $\tau \in T$ such that $\pi(y, -\tau) \in U$, or $y \in \pi(U, \tau) \subset V$. Finally if $E$ is both open and closed it is a component of $X$. QED.

VI.8. COROLLARY. *If $E$ is a compact, minimal set in a flow on a Banach space $X$, then $E$ is nowhere dense in $X$.*

*Proof:* If $E$ were dense in some nonempty open set $U$, then $U \subset \mathrm{Cl}\, U \subset E$, since $E$ is a closed set. By Theorem VI.7 we would then have $E = X$ and this would contradict the fact that $E$ is compact. QED.

*Example 1.* Let $(\phi, \theta)$, $0 \leq \phi < 1$, $0 \leq \theta < 1$, denote the coordinates of the torus $T^2$ and let

$$\frac{d\phi}{dt} = 1, \qquad \frac{d\theta}{dt} = \alpha$$

where $\alpha$ is irrational. Since every trajectory on $T^2$ is dense, it follows that $H(x) = T^2 = \Omega_x = A_x$ for every x in $T^2$. This means that $T^2$ is a compact, minimal set in this flow.

*Example 2.* On $T^2$ again define

$$\frac{d\phi}{dt} = F(\phi, \theta), \qquad \frac{d\theta}{dt} = \alpha F(\phi, \theta)$$

where $F$ is a $C^1$ function that is 1-periodic in each variable and $\alpha$ is an irrational number. Assume that $F$ is positive everywhere except at $(0, 0)$ and that $F(0, 0) = 0$. The flow on $T^2$ can be studied by considering the flow in the square $\{(\phi, \theta) : 0 \leq \phi \leq 1, 0 \leq \theta \leq 1\}$ where the points $(\phi, 1)$ and $(1, \theta)$ are identified with the points $(\phi, 0)$ and $(0, \theta)$, respectively.

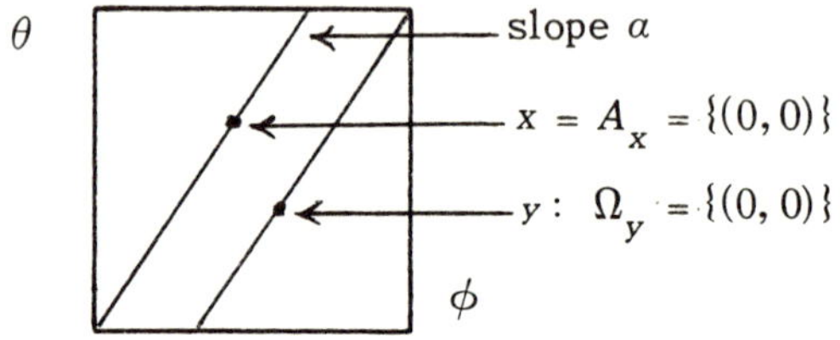


Figure 1

Since $d\theta/d\phi = \alpha$ for $(\phi, \theta) \neq (0, 0)$, the solution curves of this motion agree, in general, with those of Example 1. The single trajectory in Example 1 that passed through $(0, 0)$ is now decomposed into three trajectories, which are illustrated in Figure 1. These consist of

(i) the rest point $\{(0, 0)\}$,

(ii) those points $x$ for which $A_x = \{(0,0)\}$,

(iii) those points $y$ for which $\Omega_y = \{(0, 0)\}$.

All trajectories, other than (ii) and (iii) are Poisson stable. If $A_x = \{(0,0)\}$ and $x \neq (0, 0)$, then $x$ is not stable $P^-$. However, $x$ is stable $P^+$. The reason for this is that the positive trajectory $\gamma^+(x)$ in this flow agrees with the positive trajectory in the flow in Example 1. Hence $x \in \Omega_x$ and $x$ is stable $P^+$.

VI.C. **Almost Periodic Motions.** Let $(X, A, V)$ be a uniform space and let $\pi$ be a flow on $X$. A motion $\pi(x, t)$ is said to be **almost periodic** if the function $\pi(x, \cdot): T \to X$ is almost periodic. In this case, the point $x$ is said to be an **almost periodic point.**

Our objective here is to give a characterization of the hull of an almost periodic point. We shall give two characterizations, the first in terms of Lyapunov stability, or equicontinuity, and the second in terms of a group theoretic structure.

We shall assume throughout this section that $X$ is a complete uniform space.

If we recall the definition of almost periodicity from Chapter V we see that a motion $\pi(x, t)$ is almost periodic if and only if the set

$$\{\tau \in T: \ \pi(x, t+\tau) \in V_a(\pi(x, t)) \text{ for all } t \text{ in } T\}$$

is relatively dense in $T$ for every index $a$ in $A$. It follows immediately that every almost periodic motion is recurrent. Furthermore, it follows from Corollary V.6 that every almost periodic motion $\pi(x, t)$ is compact. Hence by Birkhoff's Recurrence Theorem, the hull $H(x)$ is a compact, minimal set.

If a compact minimal set $E$ is the hull of an almost periodic point, then we shall say that $E$ is **a.p. minimal.** If $E$ is the hull of a periodic point, then we shall say that $E$ is **periodic-minimal.** We note here, that there are examples (cf. [2]) of compact sets, even manifolds, that are minimal under some flow, but never a.p. minimal.

VI.9. LEMMA. *Let $E$ be an a.p. minimal set for a flow $\pi$ on a complete space $X$. Then every point $y \in E$ is an almost periodic point.*

*Proof:* Let $E = H(x)$ where $x$ is an almost periodic point and let $y \in H(x)$. Let $\{t_n\}$ be a sequence in $T$ with $\pi(x, t_n) \to y$. Then by Besicovitch's Theorem (Theorem V.5) we know that there is a subsequence of $\{\pi(x, t_n + t)\}$ that converges uniformly, and by the continuity of $\pi$ the limit must be $\pi(y, t)$. Since $\pi(y, t)$ is the uniform limit of almost periodic functions, it is an almost periodic function by Theorem V.1. QED.

Let us now turn to the first characterization. For this we shall use the notation $\pi_t(x) = \pi(x, t)$. Assume now that every motion $\pi(x, t)$ is defined for all $t$, that is $\pi$ is a "global" flow. It was shown in Proposition II.1 that for each $t$, the mapping $\pi_t$ is a homeomorphism of $X$ onto itself. Therefore $\pi_t$ is an element of the space $C(X, X)$ of continuous functions from $X$ into $X$. If $E$ is an invariant set in $X$, then by restricting $\pi_t$ to $E$ we see that $\pi_t \in C(E, E)$.

We shall say that the flow $\pi$ is **equi-continuous** if the family of mappings $\{\pi_t: t \in T\}$ is equi-continuous. This means that for every index $a$ and every $x \in X$, there is an index $b$ such that

$$\pi_t(y) \in V_a(\pi_t(x)) \qquad (t \in T)$$

whenever $y \in V_b(x)$. If $b$ can be chosen independent of $x$, then we shall say that the flow is **uniformly equi-continuous.** We shall say that the flow is **equi-continuous on** $E$, where $E$ is an invariant set in $X$, if the family of functions $\{\pi_t: t \in T\}$ is equi-continuous in $C(E, E)$.

The proof of the next assertion is an elementary application of compactness, and we leave it as an exercise.

VI.10. LEMMA. *Let $\pi$ be an equi-continuous flow on a compact uniform space $X$. Then $\pi$ is uniformly equi-continuous.*

VI.11. THEOREM. *Let $E$ be a compact minimal set in $X$. The set $E$ is a. p. minimal if and only if the flow $\pi$ is equi-continuous on $E$.*

*Proof:* First sssume that $E$ is a. p. minimal and let $E = H(x)$ where $x$ is an almost periodic point. We shall first show that the flow $\pi$ is equi-continuous on the trajectory $\gamma(x)$. Let an index $a$ be given and choose $b$ by Proposition I.4 so that

$$(3) \qquad x_i \in V_b(x_{i+1}), \quad i = 1, 2, 3, \Longrightarrow x_1 \in V_a(x_4).$$

Let $\ell > 0$ be an inclusion interval for the relatively dense set $E(b, \pi(x, \cdot))$. Since the continuity of $\pi$ is uniform on the compact set $E \times [0, \ell]$, there is an index $c$ such that

$$(4) \qquad \pi(y, t) \in V_b(\pi(z, t)) \qquad (0 \leq t \leq \ell),$$

whenever $y \in V_c(z)$ and $y, z \in E$.

We shall now show that

$$(5) \qquad \pi(x, t_1 + t) \in V_a(\pi(x, t_2 + t)) \qquad (t \in T)$$

whenever $\pi(x, t_1) \in V_c(\pi(x, t_2))$, which means that $\pi$ is equi-continuous on $\gamma(x)$. To prove (5) we first fix $t$ in $T$. Then there is a $\tau$ in $E(b, \pi(x, \cdot))$ such that $0 \leq \tau + t \leq \ell$. It follows from the definition of $E(b, \pi(x, \cdot))$ that

$$\pi(x, \tau + t_i + t) \in V_b(\pi(x, t_i + t)), \qquad (i = 1, 2).$$

From (4) it follows that

$$\pi(x, t_1 + \tau + t) \in V_b(\pi(x, t_2 + \tau + t)).$$

By applying (3) with $x_1 = \pi(x, t_1 + t)$, $x_2 = \pi(x, t_1 + \tau + t)$, $x_3 = \pi(x, t_2 + \tau + t)$ and $x_4 = \pi(x, t_2 + t)$ we get (5), which means that $\pi$ is equi-continuous on $\gamma(x)$.

In order to show that $\pi$ is equi-continuous on $E$ we let an index $a$ be given and choose $b$ by (3). By using the equicontinuity of $\pi$ on $\gamma(x)$ we can shoose $c$ so that

(6) $$\pi(x, t_1+t) \in V_b(\pi(x, t_2+t)) \qquad (t \in T)$$

whenever $\pi(x, t_1) \in V_c(\pi(x, t_2))$. Now choose $d$ so that

(7) $$x_i \in V_d(x_{i+1}), \quad i = 1,2,3 \Longrightarrow x_1 \in V_c(x_4) .$$

We will now show that if $y$, $z$ are points in $E$ with $y \in V_d(z)$ then $\pi(y, t) \in V_a(\pi(z, t))$ for all $t$ in $T$. Since $\gamma(x)$ is dense in $E$ we can find sequences $\{t_n\}$ and $\{s_n\}$ in $T$ so that $\pi(x, s_n) \to y$ and $\pi(x, t_n) \to z$. Without any loss of generality we can assume that $\pi(x, s_n) \in V_d(y)$ and $\pi(x, t_n) \in V_d(z)$. Since $y \in V_d(z)$ one then has $\pi(x, s_n) \in V_c(\pi(x, t_n))$ for all $n$. Hence

$$\pi(x, s_n+t) \in V_b(\pi(x, t_n+t))$$

for all $n$ and all $t$ in G. It follows from Lemma VI.9. that $\pi(x, s_n+t)$ and $\pi(x, t_n+t)$ converge to $\pi(y, t)$ and $\pi(z, t)$, respectively, uniformly in $t$. Now choose an index $N$ so that

$$\pi(x, s_n+t) \in V_b(\pi(y, t)) \text{ and } \pi(x, t_n+t) \in V_b(\pi(z, t))$$

for all $n \geq N$ and all $t \in T$. By setting $x_1 = \pi(y, t)$, $x_2 = \pi(x, s_N+t)$, $x_3 = \pi(x, t_N+t)$, $x_4 = \pi(z, t)$ and by using the symmetry of $V_b$ and (3), we see that $\pi(y, t) \in V_a(\pi(z, t))$ for all $t$ in $T$.

Now assume that the flow is equi-continuous on $E$. Since $E$ is a compact, minimal set it follows from Birkhoff's Recurrence Theorem that every motion in $E$ is recurrent. Therefore for any x in $E$ and any index $b$ the set $\{\tau: \pi(x, \tau) \in V_b(x)\}$ is relatively dense in $T$. Let the index $a$ be given and choose $b$ by the equicontinuity of $\pi$. One then has

$$\{\tau: \pi(x, \tau) \in V_b(x)\} \subset \{\tau: \pi(x, \tau+t) \in V_a(\pi(x, t)) \text{ for all } t \text{ in } T\},$$

which shows that the motion $\pi(x, t)$ is almost periodic and that $E$ is a.p. minimal. QED.

A weaker form of Theorem VI.11 is sometimes useful. For this we need another concept. We say that a flow $\pi$ on $X$ is **positively equi-continuous** if for every index $a$ and every $x$ in $X$ there is an index $b$ such that $\pi(x, t) \in V_a(\pi(y, t))$ for all $t \geq 0$ whenever $x \in V_b(y)$. Positive equi-continuity on an invariant subset is defined similarly. Uniformity is defined in a similar way, and the analogue of Lemma VI.10 is valid for this weaker concept.

VI.12. THEOREM. *Let $E$ be a compact minimal set in $X$. The set $E$ is a.p. minimal if and only if the flow $\pi$ is positively equi-continuous on $E$.*

*Proof:* In view of the last theorem it suffices to show if the flow $\pi$ is positively equi-continuous on $E$ then $E$ contains an almost periodic motion.

Since $E$ is a compact, minimal set, $E = H(x)$ where $x$ is a recurrent point. We shall show that $x$ is almost periodic. Let the index $a$ be given and choose $b$ (by Proposition I.5) so that for all $y \in X$ one has

$$V_b(y) \subset \mathrm{Cl}\, V_b(y) \subset V_a(y) \quad . \tag{8}$$

From the positive equicontinuity, choose $c$ so that

$$\pi(y, t) \in V_b(\pi(z, t)) \qquad (t \geq 0)$$

whenever $y \in V_c(z)$. Since $x$ is recurrent, the set

$$B = \{\tau \in T:\ \pi(x, \tau) \in V_c(x)\}$$

is relatively dense in $T$. It follows from the equi-continuity that for any $t \geq 0$ one has

(9) $$\pi(\pi(x, t), \tau) = \pi(\pi(x, \tau), t) \in V_b(\pi(x, t))$$

whenever $\tau \in B$. Since $\gamma^+(x)$ is dense in $E$ it follows from (9) that

$$B \subset \{\tau: \pi(y, \tau) \in \mathrm{Cl}\, V_b(y) \text{ for all } y \in E\} \cdot$$

By setting $y = \pi(x, t)$, where $t$ is arbitrary, and by using (8), we see that

$$B \subset \{\tau: \pi(x, \tau + t) \in V_a(\pi(x, t)) \text{ for all } t \text{ in } T\} .$$

Hence $\pi(x, t)$ is an almost periodic motion. QED.

We shall say that a flow $\pi$ is **distal** on an invariant set $E$ if for any pair of distinct points x and $y$ in $E$ there is an index $a$ such that

$$\pi(x, t) \notin V_a(\pi(y, t)) , \qquad (t \in T).$$

VI.13. LEMMA. *If a flow $\pi$ is equi-continuous on an invariant set E, then it is distal on E. In particular the flow on an a.p. minimal set is distal.*

*Proof:* If the flow is not distal, then there exist two distinct points x and $y$ and a sequence $\{t_n\}$ in $T$ such that for every index $b$ in $A$ there is an index $N$ with

(10) $$\pi(x, t_n) \in V_b(\pi(y, t_n)) , \qquad (n \geq N).$$

Now choose the index $a$ so that $x \notin V_a(y)$. With $a$ given choose $b$ by the equicontinuity of $\pi$. One then has $\pi(\hat{x}, t) \in V_a(\pi(\hat{y}, t))$ for all $t$ in $T$, whenever $\hat{x} \in V_b(\hat{y})$. Now choose $N$ so that (10) holds. One then has

$$x = \pi(\pi(x, t_N), -t_N) \in V_a(\pi(\pi(y, t_N), -t_N)) = V_a(y),$$

which is a contradiction. QED.

We now turn to the second characterization of a.p. minimal sets.

On any trajectory $\gamma(x)$ one can define a group structure as follows: If $y, z \in \gamma(x)$ and $y = \pi(x, s)$, $z = \pi(x, t)$, then define

$$y + z = \pi(x, s+t) \quad .$$

We shall say that this is the **natural group structure on** $\gamma(x)$. The trajectory $\gamma(x)$ then becomes an Abelian group under this natural group structure, and the identity is the point x itself.

VI.14. THEOREM. *Let $E$ be a compact minimal set in $X$. The set $E$ is a.p. minimal if and only if $E$ is an Abelian topological group with the property that there exists a trajectory $\gamma(x)$ that is a subgroup of $E$ and the inherited group structure on $\gamma(x)$ is given by the natural group structure on $\gamma(x)$.*

Before we prove this, let us make some observations. First it should be clear that the trajectory $\gamma(x)$ is determined by the fact that the identity for $\gamma(x)$ and the identity for $E$ must be the same, viz., the point x. Secondly, we shall show in the proof that the flow $\pi$ and the group operation + are related by

$$\pi(y, t) = y + \pi(x, t)$$

for all $y \in E$ and $t \in T$, where x is the identity for the group.

*Proof:* First assume that $E$ is an a.p. minimal set and let $x \in E$. Let $\gamma(x)$ have the natural group structure where x is the identity. We want to show that the Abelian group on $\gamma(x)$ can be extended to all of $E$ in such a way that $E$ is an Abelian topological group. The extension is given as follows:

Let $y$ and $z$ be two points in $E$, then there are sequences $\{s_n\}$ and $\{t_n\}$ in $T$ such that $y = \lim \pi(x, s_n)$ and $z = \lim \pi(x, t_n)$. Now define

$$y + z = \lim \pi(x, t_n + s_n) . \tag{11}$$

In order to show that $y + z$ is well-defined, we must show that $\{\pi(x, t_n + s_n)\}$ is a Cauchy sequence and that the limit of this sequence does not depend on $\{s_n\}$ and $\{t_n\}$, but only on $y$ and $z$. Since $\{\pi(x, t_n)\}$ is a convergent sequence and since the motion $\pi(x, t)$ is almost periodic, it follows from Corollary V.7 that the sequence $\{\pi(x, t_n + t)\}$ is uniformly convergent. Similarly the sequence of functions $\{\pi(x, s_n + t)\}$ is uniformly convergent. Let the index $a$ be given and choose $b$ by Proposition I.4 so that

$$x_i \in V_b(x_{i+1}), \quad i = 1,2, \longrightarrow x_1 \in V_a(x_3) . \tag{12}$$

From what was noted above, we can find an index $N$ so that

$$\pi(x, s_n + t) \in V_b(\pi(x, s_m + t)) , \qquad (n, m \geq N, \ t \in T) \tag{13}$$

and

$$\pi(x, t_n + t) \in V_b(\pi(x, t_m + t)) , \qquad (n, m \geq N, \ t \in T). \tag{14}$$

By setting $t = t_n$ in (13) and $t = s_m$ in (14) and by using (12), we get $\pi(x, t_n + s_n) \in V_a(\pi(x, t_m + s_m))$ for all $n, m \geq N$. This shows that $\{\pi(x, t_n + s_n)\}$ is a Cauchy sequence.

Next we show that $\lim \pi(x, t_n + s_n)$ depends only on $y$ and $z$. Let $\{s_n'\}$ and $\{t_n'\}$ be two other sequences with $y = \lim \pi(x, s_n')$ and $z = \lim \pi(x, t_n')$. Let

$$w = \lim \pi(x, s_n' + t_n') \ .$$

We want to show that for every index $a$ one has $w \in V_a(y + z)$ where $y + z$ is given by (11). It will follow then that $w = y + z$. It will suffice to show that for every index $a$ there is an index $N$ such that

$$\pi(x, s_n + t_n) \in V_a(\pi(x, s_n' + t_n')) \qquad (n \geq N). \tag{15}$$

It follows from Corollary V.7 that both sequences $\{\pi(x, s_n + t)\}$ and $\{\pi(x, s_n' + t)\}$ converge uniformly to $\pi(y, t)$, and $\{\pi(x, t_n + t)\}$ and $\{\pi(x, t_n' + t)\}$ converge uniformly to $\pi(z, t)$. If the index $b$ is given by (12) then we can find an index $N$ such that

$$\pi(x, s_n + t) \in V_b(\pi(x, s_n' + t)) \qquad (n \geq N, \ t \in T) \tag{16}$$

$$\pi(x, t_n + t) \in V_b(\pi(x, t_n' + t)) \qquad (n \geq N, \ t \in T). \tag{17}$$

By setting $t = t_n$ in (16) and $t = s_n'$ in (17) and by using (12) one then gets (15). This shows that $y + z$ is well-defined by (11).

It is clear that (11) defines a group on $E$, and since $t_n + s_n = s_n + t_n$, it follows that the group is commutative. We now want to show that $E$ is a topological group. That is, we want to show that the operations of addition and inversion are continuous.

Let the index $a$ be given and choose $b$ so that

$$x_i \in V_b(x_{i+1}), \quad i = 1,2,3 \Longrightarrow x_1 \in V_a(x_4) \ .$$

With $b$ given choose $c$ by the uniform equi-continuity (Theorem VI.11) so that

$$\pi(x_1, t) \in V_b(\pi(x_2, t)) \ , \qquad (t \in T),$$

whenever $x_1 \in V_c(x_2)$. Then choose the index $d$ so that

$$x_i \in V_d(x_{i+1}), \ i = 1,2,3 \implies x_1 \in V_c(x_4),$$

We will now show that

$$-\hat{y} \in V_a(-y) \quad \text{and} \quad \hat{y} + z \in V_a(y+z)$$

whenever $\hat{y} \in V_d(y)$ and $z$ is any point in $E$. Assume that $\hat{y} \in V_d(y)$ and $z \in E$. Let $\{\hat{t}_n\}$, $\{t_n\}$ and $\{s_n\}$ be sequences in $T$ with $\hat{y} = \lim \pi(x, \hat{t}_n)$, $y = \lim \pi(x, t_n)$ and $z = \lim \pi(x, s_n)$. Choose the index $N$ so that for $n \geq N$ one has

$$\pi(x, \hat{t}_n) \in V_d(\hat{y}), \quad \pi(x, t_n) \in V_d(y) \ , \tag{18}$$

$$\pi(x, -\hat{t}_n) \in V_b(-\hat{y}), \ \pi(x, -t_n) \in V_b(-y) \tag{19}$$

$$\pi(x, \hat{t}_n + s_n) \in V_b(\hat{y} + z), \quad \pi(x, t_n + s_n) \in V_b(y+z). \tag{20}$$

Inequality (18) implies that $\pi(x, t_n) \in V_c(\pi(x, \hat{t}_n))$ for $n \geq N$ and therefore

$$\pi(x, t_n + t) \in V_b(\pi(x, \hat{t}_n + t)), \qquad (t \in T), \tag{21}$$

for all $n \geq N$, by the equicontinuity of $\pi$. In particular, if we set $t = -t_n - \hat{t}_n$ in (21) and apply (19) we get $-\hat{y} \in V_a(-y)$. If we set $t = s_n$ in(21) and apply (20) we get $\hat{y} + z \in V_a(y + z)$. This completes the first half of the proof.

Let us now prove the converse. Assume that there is a group structure on the compact minimal set $E$ so that, for some trajectory $\gamma(x)$, the inherited group structure is the natural group structure on $\gamma(x)$, where $x$ is the identity of the group $E$. We will now show that the point $x$ is almost periodic.

First we claim that for every point $y$ in $E$ one has

$$\pi(y, t) = y + \pi(x, t) \ , \qquad (t \in T). \tag{22}$$

Equation (22) holds for $y \in \gamma(x)$, because $y = \pi(x, \tau)$ implies that $y + \pi(x, t) = \pi(x, \tau) + \pi(x, t) = \pi(x, \tau + t) = \pi(\pi(x, \tau), t) = \pi(y, t)$. Since $\gamma(x)$ is dense in $E$ and since addition is continuous on $E$, it follows that (22) holds for all $y$ in $E$.

Since $E$ is compact, addition is uniformly continuous on $E$. This means that for every index $a$ there is an index $b$ such that for any $z$ in $E$ one has $\hat{y} + z \in V_a(y + z)$ whenever $\hat{y} \in V_b(y)$. If we set $z = \pi(x, t)$ and use (22) this becomes

$$\pi(\hat{y}, t) \in V_a(\pi(y, t)), \qquad (t \in T), \tag{23}$$

whenever $\hat{y} \in V_b(y)$. Since $\pi(x, t)$ is recurrent (by Birkhoff's Recurrence Theorem), the set $\{\tau\colon \pi(x, \tau) \in V_b(x)\}$ is relatively dense in $T$. However, (23) implies that

$$\{\tau\colon \pi(x, \tau) \in V_b(x)\} \subset E(a, \pi(x, \cdot))$$

and hence the point $x$ is almost periodic in the flow $\pi$. QED.

The concept of equicontinuity is related to the concept of Lyapunov stability. This relationship will be used when we study solutions of differential equations.

We shall say that a motion $\pi(x, t)$ is **Lyapunov-stable (with respect to a set** $D$) if $\pi(x, t)$ is defined for all $t$ and for every index $a$ there is an index $b$ such that

$$\pi(y, t) \in V_a(\pi(x, t)), \qquad (t \in T), \tag{24}$$

whenever $y \in D$ and $y \in V_b(x)$. The motion $\pi(x, t)$ is said to be **uniformly Lyapunov-stable (with respect to** $D$) if $\pi(x, t)$ is defined by all $t$ and for every index $a$ there is an index $b$ such that

$$(25) \qquad \pi(y, t) \in V_a(\pi(\hat{x}, t)), \qquad (t \in T),$$

whenever $y \in D$, $\hat{x} \in \gamma(x)$ and $y \in V_b(\hat{x})$. We define **positive Lyapunov stability** by requiring $\pi(x, t)$ to be defined only for $t \geq 0$ and (24) to hold for $t \geq 0$. Similarly, **uniform, positive Lyapunov stability** is defined by requiring (25) to hold for $t \geq 0$ while $\hat{x} \in \gamma^+(x)$. (Negative stabilities are defined in an analogous way.) If $D = X$, then we shall drop the phrase "with respect to $D$."

The pertinent relationships between almost periodicity, equicontinuity and Lyapunov stability are given in the following statements. We shall leave the proofs as exercises.

1. The motion $\pi(x, t)$ is uniformly Lyapunov stable with respect to $\gamma(x)$ if and only if the flow is equicontinuous on $\gamma(x)$.

2. If every motion $\pi(x, t)$ in a compact invariant set $E$ is Lyapunov-stable with respect to $D$ where $E \subset D$, then every motion is uniformly Lyapunov-stable with respect to $D$.

3. A compact motion $\pi(x, t)$ is almost periodic if and only if it is recurrent and uniformly positively Lyapunov-stable with respect to $\gamma(x)$.

4. If a motion $\pi(x, t)$ is negatively compact and uniformly positively Lyapunov-stable with respect to $\gamma(x)$, then $\pi(x, t)$ is almost periodic.

VI.D. **Kakutani-Baum Theorem.** In this section we shall give another characterization of almost periodic flows. The very definition of almost periodicity depends on the uniform structure on a space $X$. Thus almost periodicity differs from the concept of recurrence, which is a topological concept. One of the consequences of this characterization due to Kakutani and Baum [4] is that, to a limited extent, almost periodicity is independent of the uniform structure, cf. Theorem VI.18 and its corollary.

Let $\pi$ be a flow on a compact, uniform space $X$, and assume that $X$ is a minimal set. Let $B$ be any Banach space with norm $\|\cdot\|$ and consider the space $C(X, B)$ of continuous functions from $X$ into $B$. For each $f \in C(X, B)$ and each $x \in X$, the function $f(\pi(x, \cdot))$ maps $T$ into $B$. The theorem we wish to prove is the following:

VI.15 THEOREM. (Kakutani-Baum) *In the above notation, the set $X$ is a.p. minimal if and only if there is a point $x_0 \in X$ such that for every continuous function $f\colon X \to B$, the function*

$$f(\pi(x_0, \cdot))\colon T \to B$$

*is Bohr almost periodic.*

It is clear that the above condition is necessary in order that $X$ be a.p. minimal; the problem is to show the sufficiency. The first lemma shows that the point $x_0$ can be replaced by any other point $x$ in $X$.

VI.16 LEMMA. *Let $\pi$ be a flow on a compact uniform space $X$ and assume that $X$ is minimal. Assume that there is an $x_0 \in X$ such that for every continuous function $f$ in $C(X, B)$, the function $f(\pi(x_0, \cdot))$ is Bohr almost periodic. Then $f(\pi(x, \cdot))$ is Bohr almost periodic for every $x$ in $X$ and moreover, for every $\varepsilon > 0$ the set*

$$B(f) = \{\tau \in T\colon \|f(\pi(x, \tau + t)) - f(\pi(x, t))\| < \varepsilon, \text{ for all } x \in X,\ t \in T\}$$

*is relatively dense in $T$.*

*Proof:* First we note that the set

$$D = E(\varepsilon/3, f(\pi(x_0, \cdot))),$$

is relatively dense in $G$. Since $f$ is uniformly continuous one can find an index $\alpha$ so that

$$(26) \qquad y \in V_\alpha(x) \Longrightarrow \|f(x) - f(y)\| < \varepsilon/3 \qquad (\text{all } x \in X).$$

Now choose $\tau \in D$ and $t \in T$. Since $\pi(\cdot . t)$ and $\pi(\cdot, \tau + t)$ are uniformly continuous one can find an index $\beta$ so that

$$(27) \qquad \begin{aligned} y \in V_\beta(x) \Longrightarrow \pi(y, t) &\in V_\alpha(\pi(x, t)) \text{ and} \\ \pi(y, \tau + t) &\in V_\alpha(\pi(x, \tau + t)) \end{aligned}$$

for all $x$ in $X$. Since $X$ is minimal one can find a $t_1$ in $T$ so that $\pi(x_0, t_1) \in V_\beta(x)$. It now follows from (26) and (27) that

$$\|f(\pi(x, \tau + t)) - f(\pi(x_0, t_1 + \tau + t))\| < \varepsilon/3$$

$$\|f(\pi(x, t)) - f(\pi(x_0, t_1 + t))\| < \varepsilon/3$$

Since $\|f(\pi(x_0, \tau + t_1 + t)) - f(\pi(x_0, t_1 + t))\| < \varepsilon/3$ because $\tau \in D$, it follows that

$$\begin{aligned} \|f(\pi(x, \tau+t)) - f(\pi(x, t))\| \le\ & \|f(\pi(x, \tau+t)) - f(\pi(x_0, t_1+\tau+t))\| \\ & + \|f(\pi(x_0, t_1+\tau+t)) - f(\pi(x_0, t_1+t))\| \\ & + \|f(\pi(x_0, t_1+t))\| - f(\pi(x, t))\| \le \varepsilon \end{aligned}$$

that is $D \subset B(f)$. QED.

Another lemma we need is a direct consequence of Urysohn's lemma (cf. [15, p 115]). We shall omit the proof.

VI.17. LEMMA. *Let $X$ be a compact, uniform space. Then for each index $\alpha$ there exists a finite collection of $B$-valued functions $\{f_1, \ldots, f_n\}$ such that $x \in V_\alpha(y)$ whenever $\|f_i(x) - f_i(y)\| < 1/2$ for $1 \leq i \leq n$.*

*Proof of Kakutani-Baum Theorem:* Let $\alpha$ be any index for $X$ and choose the functions $\{f_1, \ldots, f_n\}$ by Lemma VI.17. With $\varepsilon = \frac{1}{2}$ choose $B(f_1), \ldots, B(f_n)$ by Lemma VI.16. It follows from Lemma V.8 that the set

$$D = B(f_1) \cap \cdots \cap B(f_n)$$

is relatively dense in $T$. Thus if $\tau \in D$ one has

$$\|f_i(\pi(x, \tau + t)) - f_i(\pi(x, t))\| < \tfrac{1}{2} \qquad (1 \leq i \leq n)$$

for all $x \in X$ and all $t$ in $T$. It follows then from Lemma VI.17 that $\pi(x, \tau + t) \in V_\alpha(\pi(x, t))$ for all $x \in X$ and $t$ in $T$. Hence the set $X$ is a.p. minimal QED.

Let us give an application of the Kakutani-Baum Theorem. For this application, the set $X$ will be assumed to be an a.p. minimal set set in a flow $\pi$. Specifically, we shall assume that $X$ is a space of functions $f$ mapping $T$ into a Banach space $B$, and that the flow $\pi$ is given by the translation mapping $\pi(f, t) = f_t$.

We shall let $\Delta_\tau$ denote the evaluation mapping, that is $\Delta_\tau$ maps the function $f$ into the vector $f(\tau)$, where $\tau \in T$.

VI.18. THEOREM. *Let $X$ be a collection of functions $f\colon T \to B$ where $B$ is a Banach space and assume that $X$ has a topology $\mathfrak{T}$. Assume that $X$ is an a.p. minimal set in the flow $\pi(f, t) = f_t$. If for some $\tau_0$, the evaluation mapping $\Delta_{\tau_0}\colon X \to B$ is continuous, then the functions $f$ in $X$ are Bohr almost periodic.*

*Proof:* We note that

$$\Delta_{\tau_0}(\pi(f,t)) = \Delta_{\tau_0}(f_t) = f_t(\tau_0) = f(t+\tau_0),$$

and the result now follows from the Kakutani-Baum Theorem. QED.

VI.19. COROLLARY. *Let $f_t$ be an almost periodic motion in the flow on $C(R, R^n)$ constructed in Example III.E. Then $f$ is Bohr almost periodic.*

*Proof:* We will show that the evaluation mapping

$$\Delta: C(R, R^n) \to R^n,$$

given by $\Delta: f \to f(0)$, is continuous. Since $f_n \to f$ in $C(R, R^n)$ means that $f_n$ converges to $f$ uniformly on every compact set in $R$. In particular $f_n(0) \to f(0)$. Hence $\Delta$ is continuous and the proof now follows. QED.

*Exercises 1.* Let $\pi(x, t)$ be a continuous flow on a uniform space $X$ and let $\hat{\pi}(x, t)$ be an induced discrete flow on $X$. Show that a point $x$ in $X$ is stable $P^+$ for $\pi$ if and only if it is stable $P^+$ for $\hat{\pi}$.

2. An invariant set $E$ is said to be **decomposable** if there exist two closed invariant proper subsets $A$, $B$ such that $E = A \cup B$. Otherwise $E$ is **indecomposable.** A trajectory $\pi(x, t)$ is said to be **isolated** in an invariant set $E$ if $\gamma(x)$ is an open set in $E$. An invariant set $E$ is said to be **quasi-minimal** if it is indecomposable and contains no isolated trajectories. Show that a set $E$ is quasi-minimal if and only if $E = H(x)$, where $x$ is a nonperiodic Poisson stable point (For other characterizations of quasi-minimal sets, cf [21].)

*3.* Show that there exist compact, recurrent motions that are not almost periodic. (Cf. [3].)

*4.* Let $\pi$ be a equicontinuous flow on a space $X$. Show that $\pi$ can be extended (uniquely) to a flow $\hat{\pi}$ on the completion $\hat{X}$. What happens if $\pi$ is not equicontinuous?

*5.* Let $f\colon R \to R^n$ be a bounded, uniformly continuous function and assume that the flow $f_t$ is almost periodic in the flow on $\mathfrak{L}_p(R, R^n)$ in Example III.L. Show that $f$ is Bohr almost periodic. (It is an unresolved question as to whether this conclusion remains true if one only assumes $f$ to be continuous.)

# CHAPTER VII

## THE STRUCTURE OF $\omega$-LIMIT SETS

Let $\pi$ be a flow on a uniform space $X$ and let $\pi(x, t)$ be a positively compact motion. We know then (cf. Theorem II.8) that the $\omega$-limit set $\Omega_x$ is nonempty, compact and invariant. In this section we shall try to find conditions (necessary and/or sufficient) that

(i) $\Omega_x$ be a minimal set, or that

(ii) $\Omega_x$ be an a.p. minimal set, or that

(iii) $\Omega_x$ be a periodic minimal set.

The conditions we give will be topological and usually formulated in terms of stability. Later we shall apply these results to demonstrate the existence of recurrent, almost periodic or periodic solutions of differential equations.

VII.A. **Minimal case.** We shall say that a motion $\pi(x, t)$ **approximates $\Omega_x$ uniformly** if for every index $\alpha$ there is a $T > 0$ such that $\Omega_x \subset V_\alpha(L)$ for each arc $L$ of $\gamma^+(x)$ whose (time) length is greater than $T$. (The set $V_\alpha(L)$ is, of course, $\cup_{y \in L} V_\alpha(y)$.)

VII.1. THEOREM. *Let $\pi(x, t)$ be a positively compact motion. A necessary and sufficient condition that $\Omega_x$ be a minimal set is that $\pi(x, t)$ approximates $\Omega_x$ uniformly.*

Since we will not use this result in the sequel we shall not give a proof here. We refer the reader to [22, p. 401] for a proof in the case that $X$ is a metric space.

VII.B. **A.P. Minimal Case.** The above condition that $\Omega_x$ be minimal is not easy to verify, perhaps because it is both necessary and sufficient. We next give a sufficient condition that $\Omega_x$ be a.p. minimal. This condition appears to be useful since it can be easily checked. We shall see later that it is not a necessary condition.

VII.2. THEOREM. *Let $\pi(x, t)$ be a positively compact motion and assume that $\pi(x, t)$ is uniformly positively Lyapunov-stable with respect to $\Omega_x$. Then $\Omega_x$ is an a.p. minimal set.*

*Proof:* We will first show that $\Omega_x$ is a minimal set. In order to do this it will suffice to show that for every pair of points $y$ and $z$ in $\Omega_x$ and every index $a$ there is a $\tau$ in $T$ such that $\pi(y, \tau) \in V_a(z)$.

With $a$ given, choose the index $b$ so that

$$(1) \qquad x_i \in V_b(x_{i+1}), \quad i = 1,2 \Longrightarrow x_1 \in V_a(x_3) \ .$$

And with $b$ so chosen, choose $c$ by the Lyapunov-stability so that

$$(2) \qquad \pi(y, t) \in V_b(\pi(x, \sigma + t)) \qquad (t \geq 0)$$

Provided $y \in V_c(\pi(x, \sigma))$, for $\sigma \geq 0$, and $y \in \Omega_x$. Now choose sequences $\{s_n\}$ and $\{t_n\}$ in $T^+$ so that $s_n \to \infty$, $t_n \to \infty$ and

$$x_n = \pi(x, s_n) \to y, \quad \hat{x}_n = \pi(x, t_n) \to z \ .$$

Fix the index $m$ so that $x_m = \pi(x, s_m) \in V_c(y)$, or $y \in V_c(x_m)$. Then

$$(3) \qquad \hat{x}_n = \pi(x, t_n) = \pi(x_m, t_n - s_m)$$

for all $n$. Since $t_n \to \infty$, there is an index $N_1$ with $t_n - s_m \geq 0$ for all $n \geq N_1$. From (2) and (3) this implies that

$$\pi(y, t_n - s_m) \in V_b(\hat{x}_n), \qquad (n \geq N_1).$$

Now choose the index $N$ so that $N \geq N_1$ and $\hat{x}_n \in V_b(z)$ for $n \geq N$. Let $\tau = t_N - s_m$, then

$$\pi(y, \tau) \in V_b(\hat{x}_N), \quad \hat{x}_N \in V_b(z) \implies \pi(y, \tau) \in V_a(z),$$

which shows that $\Omega_x$ is a minimal set.

In order to show that $\Omega_x$ is a.p. minimal, we will show that the flow $\pi$ is positively equi-continuous on $\Omega_x$ and then apply Theorem VI.12. Let the index $a$ be given and choose $b$ so that (1) holds. Choose $c$ so that (2) holds. Now choose the index $d$ so that $d \geq c$ and

$$x_i \in V_a(x_{i+1}), \ i = 1, 2, \implies x_1 \in V_c(x_3).$$

Let $y$ and $z$ be any two points in $\Omega_x$ with $y \in V_a(z)$. One can then find a point $x_1 \in \gamma^+(x)$ such that $x_1 \in V_d(y) \subset V_c(y)$. Hence $x_1 \in V_c(z)$. It follows then from (2) that

$$\pi(x_1, t) \in V_b(\pi(y, t)) \text{ and } \pi(x_1, t) \in V_b(\pi(z, t)), \qquad (t \geq 0).$$

Then from (1) we get $\pi(y, t) \in V_a(\pi(z, t))$ for all $t \geq 0$, which shows that $\pi$ is positively equi-continuous on $\Omega_x$. QED.

VII.3 COROLLARY. *Let $\pi(x, t)$ be a positively compact motion that is uniformly positively Lyapunov-stable with respect to $\Omega_x$. Then every motion in $\Omega_x$ is uniformly Lyapunov-stable with respect to $\Omega_x$.*

*Proof.* Since $\Omega_x$ is an a.p. minimal set, this follows from Theorem VI.11. QED.

VII.C. **Periodic Minimal Case. (Limit Cycle-Limit Point).** In order to show that an $\omega$-limit set is periodic minimal, a stronger form of stability is needed.

We shall say that a motion $\pi(x, t)$ is **asymptotically stable with respect to** $D$ if (i) it is uniformly positively Lyapunov-stable with respect to $D$ and (ii) there is an index $a$ such that for any $y \in V_a(\gamma^+(x))$ there is a $\tau$ in $T$ with the following property: for every index $b$, there is a $K \geq 0$ such that

$$\pi(y, t) \in V_b(\pi(x, \tau+t)) \text{ for } t \geq K.$$

(In a metric space, (4) can be reformulated by saying that

$$\rho(\pi(y, t), \pi(x, \tau+t)) \to 0 \text{ as } t \to \infty$$

where $\rho$ denotes the metric.)

Property (ii) above can be formulated another way. Let $A(x)$ denote the **region of attraction** for $\gamma^+(x)$, that is $A(x)$ consists of all points $y$ in $X$ for which there is a $\tau$ in $T$ such that (4) holds. Property (ii) then is equivalent to saying that $D \cap V_a(\gamma^+(x)) \subset A(x)$.

Note that if $D$ is closed, then

$$D \cap V_a(\gamma^+(x)) = D \cap V_a(H^+(x))$$

where $H^+(x) = \text{Cl } \gamma^+(x)$. Also note that if $\pi(x, t)$ is asymptotically stable with respect to the closed set $\Omega_x$, then for some index $a$ one has

$$\Omega_x \subset \Omega_x \cap V_a(H^+(x)) \subset A(x) . \tag{5}$$

VII.4 THEOREM. *Let $\pi(x, t)$ be a positively compact motion and assume that $\pi(x, t)$ is asymptotically stable with respect to $\Omega_x$. Then $\Omega_x$ is a periodic minimal set.*

*Proof:* First we claim that for any two points $y$ and $z$ in $\Omega_x$, there is $\rho$ in $T$ such that for every index $b$ there is a $K \geq 0$ such that $\pi(z, t) \in V_b(\pi(y, \rho + t))$ for all $t \geq K$. In order to prove this we first note that it follows from (5) that $\Omega_x \subset A(x)$. Hence there exist times $\sigma$ and $\tau$ in $T$ such that for any index c, there is $K \geq 0$ such that

$$\pi(y, t) \in V_c(\pi(x, \sigma + t)) \text{ and } \pi(z, t) \in V_c(\pi(x, \tau + t))$$

for all $t \geq K$. It follows then that

$$\text{(6)} \qquad \pi(z, t) \in V_c(\pi(x, \tau + t)) \text{ and } \pi(x, \tau + t) \in V_c(\pi(y, \tau - \sigma + t))$$

for all $t \geq \max(K, K + \sigma - \tau)$. If c is now chosen so that

$$x_i \in V_c(x_{i+1}), \ i = 1,2 \Longrightarrow x_1 \in V_b(x_s),$$

we see that (6) implies that

$$\text{(7)} \qquad \pi(z, t) \in V_b(\pi(y, \tau - \sigma + t))$$

for all $t \geq \max(K, K + \sigma - \tau)$.

It follows from Theorem VII.2 that $\Omega_x$ is an a.p. minimal set. Consequently, Lemma VI.13 implies that the flow is distal on $\Omega_x$. We now claim that there is some $y$ in $\Omega_x$ for which $\Omega_x = \gamma(y)$. If this were not true, then for any $y$ in $\Omega_x$ one can find a $z$ in $\Omega_x$, $z \notin \gamma(y)$. If we now apply (7) and note that $z \neq \pi(y, \tau - \sigma)$, we see that the flow is not distal on $\Omega_x$, which is a contradiction. Since $H(y) = \Omega_x = \gamma(y)$, for some $y$ in $\Omega_x$, it follows from Corollary VI.4 that the motion $\pi(y, t)$ is periodic and that $\Omega_x$ is a periodic minimal set. Q.E.D.

The following corollary is almost immediate.

VII.5. COROLLARY. *Let $\pi(x, t)$ be a positively compact motion that is asymptotically stable with respect to $\Omega_x$. Then every motion $\pi(y, t)$ in $\Omega_x$ is asymptotically stable with respect to $\Omega_x$.*

The last theorem gives a topological-dynamical description of the classical "limit cycle" case, cf. [8]. We now turn our attention to the "limit point" case, that is the case where $\Omega_x$ consists of a single point. The type of stability now required is stronger than the asymptotic stability of the last theorem.

VII.6 THEOREM. *Let $\pi(x, t)$ be a positively compact motion. Then the following statements are equivalent:*

(A) *$\Omega_x$ consists of a single point.*

(B) *$\pi(x, t)$ is uniformly positively Lyapunov-stable with respect to $\Omega_x$ and for every $y$ in $\Omega_x$ and every index $b$ there is a $K \geq 0$ such that $\pi(x, t) \in V_b(\pi(y, t))$ for all $t \geq K$.*

*Proof:* Assume that (A) holds. Then $\pi(x, t) \to \Omega_x = \{y\}$ as $t \to \infty$. That is, for every index $b$ there is a $K \geq 0$ such that

$$\pi(x, t) \in V_b(\Omega_x) = V_b(y) = V_b(\pi(y, t))$$

for all $t \geq K$. The Lyapunov-stability is easily verified and we see that (B) holds.

Now assume that (B) holds. Then $\pi(x, t)$ is asymptotically stable with respect to $\Omega_x$ (set $\tau = 0$). Hence $\Omega_x$ is a periodic minimal set by Theorem VII.4. It follows from Lemma VI.13 that the flow is distal on $\Omega_x$. Let the index $b$ be given and choose c so that

$$x_i \in V_c(x_{i+1}),\ i = 1,2, \Longrightarrow x_1 \in V_b(x_3) .$$

Let $y \in \Omega_x$ and $\pi(y, \tau) \in \Omega_x$, where $\tau$ is a (fixed) time in $T$. It follows from (B) that there is a $K \geq 0$ such that

$$\pi(x, t) \in V_c(\pi(y, t)) \text{ and } \pi(x, t) \in V_c(\pi(y, \tau + t))$$

for all $t \geq K$. This implies that

$$\pi(y, t) \in V_b(\pi(\pi(y, \tau), t))$$

for all $t \geq K$. Since the flow is distal, this means that $y = \pi(y, \tau)$, that is (A) holds. Q.E.D.

The type of stability given in the last theorem is related to the familiar Lyapunov-asymptotic stability, which differs from our concept of asymptotic stability. For a general discussion, comparing these types of asymptotic stabilities with some types of orbital stability, see [27, pp. 155-157]. Also see [16] and [33].

VII.D. **An example.** Let us show that the condition of uniform positive Lyapunov-stability of $\pi(x, t)$ is not a necessary condition that $\Omega_x$ be an a.p. minimal set. We shall show, in fact, that it is not a necessary condition for $\Omega_x$ to be a periodic minimal set.

Using polar coordinates $(r, \theta)$ in the plane $R^2$, we consider the differential equation

$$r' = 1 - r, \quad \theta' = f(r) \tag{8}$$

where

$$f(r) = \begin{cases} 2\pi & 0 < r \leq 1 \\ \dfrac{-2\pi \log(r-1)}{1 - \log(r-1)} & 1 < r < 1 + e^{-1} \\ \pi & 1 + e^{-1} \leq r \,. \end{cases}$$

The solutions of (8) are given by

$$r(t) = r(r_0, \theta_0, t) = 1 + (r_0 - 1)e^{-t}$$

$$\theta(t) = \theta(r_0, \theta_0, t) = \theta_0 + 2\pi t + g(r_0, t)$$

where

$$g(r_0, t) = \begin{cases} 0 & 0 < r_0 \leq 1 \\ 2\pi\{\log|1-k| - \log|t+1-k|\} & 1 < r_0, \ k+1<t \\ -\pi t & 1 < r_0, \ t \leq k+1 \end{cases}$$

and $k = \log(r_0 - 1)$.

If $x_0 = (r_0, \theta_0)$ we define $\pi(x_0, t) = (r(r_0, \theta_0, t), \theta(r_0, \theta_0, t))$. For $x = (1+e^{-1}, 0)$ we have

$$\pi(x, t) = (1+e^{-1-t}, 2\pi\{t + \log 2 - \log(t+2)\}), \qquad t \geq 0.$$

Also, $\Omega_x = \{(r, \theta): r = 1\}$, which is a periodic minimal set. That is, if $y = (1, 0)$, then $y \in \Omega_x$ and $\pi(y, t) = (1, 2\pi t)$.

Now define sequences $\{s_m\}$ and $\{t_m\}$ so that

$$s_m \geq 0, \quad t_m \geq 0$$

$$s_m \to \infty, \quad t_m \to \infty$$

$$x_m = \pi(x, s_m) \to y$$

$$\rho(\pi(x_m, t_m), \ \pi(y, t_m)) \geq 2$$

where $\rho$ denotes the Euclidean metric on $R^2$. The sequences $\{s_m\}$ and $\{t_m\}$ can be defined as follows: Let $s_0 = 0$ and let $s_m$ $(1 \leq m)$ be defined by

$$s_m - \log(s_m + 2) = m - \log 2 \ .$$

Then define $t_m$ by $t_m = (s_m + 2)(\sqrt{e} - 1)$ for $0 \leq m$. One can then show that

$$\rho(\pi(x, s_m), y)^2 = \exp(-2-2s_m) \to 0$$

and

$$\rho(\pi(x_m, t_m), \pi(y, t_m))^2 = \exp(-2-2s_m - 2t_m)$$
$$- 2\{1 + \exp(-1 - s_m - t_m)\}\{\cos(2\pi(m - ½)) - 1\} \geq 4 \ .$$

It follows then that the motion $\pi(x, t)$ is not uniformly positively Lyapunov-stable with respect to $\Omega_x$ and therefore it is not asymptotically stable with respect to $\Omega_x$.

# CHAPTER VIII

## *APPLICATIONS TO DIFFERENTIAL EQUATIONS*

In this chapter we return to the study of the flow constructed in Chapter IV for nonautonomous differential equations. We want to study specifically questions of stability and questions of the existence of periodic and almost periodic solutions. Before turning our attention to these questions, we must complete an analysis of several questions first posed in Chapter IV.

Throughout this chapter we will study the solutions $\phi(x, f, t)$ of differential equations $x' = f(x, t)$ where $f \in C(W \times R, R^n)$ and $f$ satisfies the strong uniqueness condition. We will first compare the solution $\phi(x, f, t)$ with the motion $\pi(x, f, t)$, continuing the analysis started in Section IV.3. We will then consider some specific differential equations and study the corresponding set of limiting equations. After this we will then turn to the question of stability and the question of the existence of periodic and almost periodic solutions.

Recall that the topology of uniform convergence on compact sets on $C(W \times R, R^n)$ is metrizable.

VIII.A. **The Solution $\phi(x, f, t)$ and the Motion $\pi(x, f, t)$.** Consider the differential equation $x' = f(x, t)$ where $f \in C(W \times R, R^n)$ and $f$ satisfies the strong uniqueness condition, that is, $f$ is regular. As shown in Chapter IV, the mapping

$$\pi(x, f, t) = (\phi(x, f, t), f_t)$$

defines a flow on $W \times H(f)$. Let

$$P: W \times H(f) \to W \text{ and } Q: W \times H(f) \to H(f)$$

be the projections defined in Section IV.3.

We shall say that a solution $\phi(x, f, t)$ of $x' = f(x, t)$ is **positively Poisson stable** if for every neighborhood $U$ of $x$ in $W$, the set

$$\{\tau \in T: \tau \geq 0 \text{ and } \phi(x, f, \tau) \in U\}$$

is unbounded in $T$. **Negative Poisson stable** is defined similarly by requiring $\tau$ to satisfy $\tau \leq 0$ instead of $\tau \geq 0$. The solution $\phi$ is **Poisson stable** if it is both positively and negatively Poisson stable. A solution $\phi(x, f, t)$ is said to be **recurrent** if for every neighborhood $U$ of $x$ the set

$$\{\tau \in T: \phi(x, f, \tau) \in U\}$$

is relatively dense in $T$.

The next lemma is easily verified.

VIII.1 LEMMA. *Let $f$ be a regular function in $C(W \times R, R^n)$, If a motion $\pi(x, g, t)$ in $W \times H(f)$ has Property X, then the motion $g_t$ in $H(f)$ has Property X and the corresponding solution $\phi(x, g, t)$ has Property X, where X is any one of the following: Poisson-stability, recurrence, almost periodicity or periodicity.*

What about the converse question? Is it true that if $g_t$ and $\phi(x, g, t)$ have Property $X$, this implies that $\pi(x, g, t)$ has Property $X$. First we observe that this is not true for periodicity. That is, if $g_t$ and $\phi(x, g, t)$ are periodic with incommensurable periods, then $\pi(x, g, t)$ is almost periodic but not periodic. We can prove the following result however:

VIII.2 LEMMA. *Let $f$ be a regular function in $C(W \times R, R^n)$, If $f_t$ is almost periodic in $H(f)$ and the solution $\phi(x, f, t)$ is almost periodic, then the motion $\pi(x, f, t)$ is almost periodic.*

*Proof:* This is a direct application of Theorem V.9. Q.E.D.

VIII.B **Structure of limiting equations.** In this section we shall apply the conclusions of Chapter VII to the study of the set of limit equations for $x' = f(x, t)$ where $f \in C(W \times R, R^n)$,

Assume that $f_t$ is a positively compact motion in $C(W \times R, R^n)$ and let $\Omega_f$ denote the $\omega$-limit set. We shall say that $f$ is **asymptotically autonomous** if $\Omega_f$ consists of a single point. $f$ is **asymptotically periodic** if $\Omega_f$ is a periodic minimal set. $f$ is **asymptotically almost periodic** if $\Omega_f$ is an a.p. minimal set. $f$ is **asymptotically recurrent if** $\Omega_f$ is a (compact) minimal set. $f$ is **asymptotically Poisson-stable** if there is a $g$ in $\Omega_f$ such that $H(g) = \Omega_f$.

It is possible to characterize those functions $f$ in $C(W \times R, R^n)$ that are asymptotically autonomous.

VIII.3 THEOREM. *Let $f \in C(W \times R, R^n)$ and assume that the motion $f_t$ is positively compact. Then $f$ ia asymptotically autonomous if and only if there exist functions $g$ and $h$ in $C(W \times R, R^n)$ such that*

(i) $f = g + h$,

(ii) *$g$ is autonomous, that is, $g$ depends only on $x \in W$,*

(iii) *$h(x, t) \to 0$ as $t \to \infty$, uniformly on compact sets in $W$.*

Before proving this result, let us first examine condition (iii) above.

VIII.4 LEMMA. *Let* $h \in C(W \times R, R^n)$ *and assume that* $h(x, t) \to 0$ *as* $t \to \infty$, *uniformly on compact sets in* $W$. *Then* $h_t \to 0$ *as* $t \to \infty$ *in the flow on* $C(W \times R, R^n)$.

*Proof:* In order to show that $h_t \to 0$ as $t \to \infty$ we want to show that for every pair of compact sets $K \subset W$, $I \subset R$ and every $\varepsilon > 0$, there is a $T$ such that $|h_t(x, s)| \leq \varepsilon$ whenever, $s \in K$, $s \in I$ and $t \geq T$. So let $K$, $I$ and $\varepsilon$ be given. We can enlarge $I$, if necessary, so that $I$ is an interval $[-a, a]$. Since $h(x, t) \to 0$ as $t \to \infty$, we can find a $\tau$ such that $|h(x, t)| \leq \varepsilon$ whenever $x \in K$ and $t \geq \tau$. If we set $T = \tau + a$ we get the desired result. Q.E.D.

*Proof of Theorem VIII.3:* If $f = g + h$ where $g$ and $h$ satisfy (ii) and (iii), it follows from Lemma VIII.4 that $\Omega_f = \{g\}$. Hence $f$ is asymptotically autonomous.

Conversely, if $f$ is asymptotically autonomous, then $\Omega_f = \{g\}$ where $g$ is an autonomous function. Let $h = f - g$. We want to show that $h$ satisfies (iii). Since $\Omega_f = \{g\}$, this means that for every sequence $\{t_n\}$, with $t_n \to \infty$, and every pair of compact sets $K$, $I$ ($K \subset W$, $I \subset R$) one has

$$f(x, t_n + t) \to g(x)$$

uniformly for $x \in K$, $t \in I$. Now let $K$ be fixed (but arbitrary), let $I = [0, 1]$ and $t_n = n$. Then for every $\varepsilon > 0$ we can find an $N$ so that for $n \geq N$ one has

$$\sup_{x \in K,\ t \geq N} |h(x, t)| = \sup_{x \in K,\ t \geq N} |f(x, t) - g(x)|$$

$$= \sup_{x \in K,\ t \in I} |f(x, n + t) - g(x)| \leq \varepsilon \ .$$

Therefore $h$ has property (iii). Q.E.D.

Necessary conditions for asymptotic periodicity and asymptotic almost periodicity are harder to come by. This problem arose in Chapter VII. However, one can give sufficient conditions which are quite practical.

VIII.5. THEOREM. *Let $f$, $g$ and $h$ be functions in $C(W \times R, R^n)$ where $f = g + h$ and $f_t$ is a positively compact motion. Assume that $h(x,t) \to 0$ as $t \to \infty$, uniformly on compact sets in $W$. Then the following statements are valid:*

(A) *The motion $g_t$ is positively compact and $\Omega_f = \Omega_g$.*

(B) *If the motion $g_t$ is periodic {almost periodic, recurrent, Poisson-stable}, then the motion $f_t$ is asymptotically periodic {asymptotically almost periodic, asymptotically recurrent, asymptotically Poisson-stable}. Moreover, $\Omega_f = H(g)$.*

*Proof:* This follows immediately from Lemma VIII.4. Q.E.D.

The remainder of this section is a lengthy exercise. We ask the reader to fill in all the details. The results of this section can be applied to differential equations satisfying Carathéodory conditions. (Cf. Section IV.E.)

Let $f = g + h$ where $f$, $g$ and $h$ are in $\mathfrak{L}_p(W \times R, R^n)$. Assume that

$$(1) \qquad \int_t^{t+1} |h(x,s)|^p \, ds \to 0 \quad \text{as} \quad t \to \infty ,$$

uniformly on compact sets in $W$.

1. Show that if $h$ satisfies (1) then

$$h_t \to 0 \text{ as } t \to \infty$$

in the topology $\mathcal{T}_b$.

2. Show that if $h$ is continuous and satisfies

$$h(x, t) \to 0 \text{ as } t \to \infty, \tag{2}$$

uniformly on compact sets in $W$, then $h$ satisfies (1).

3. Show that if for every compact set $K \subset W$, there is a real number $B$ such that

$$\int_0^\infty |h(x, s)|^p \, ds \leq B, \qquad (x \in K) \tag{3}$$

then $h$ satisfies (1).

4. Assume that $h$ satisfies (1) and that $f_t$ is a compact motion in $(\mathfrak{L}_p, \mathcal{T}_b)$. Show that $g_t$ is compact and that $\Omega_f = \Omega_g$.

5. Assume that $g \in C(W \times R, R^n)$ and that $g$ is either periodic, almost periodic, recurrent or Poisson-stable. Assume also that $h$ satisfies (1). Show that $f$ is, respectively, asymptotically periodic, asymptotically almost periodic, asymptotically recurrent, or asymptotically Poisson-stable.

For the topology $\mathcal{T}_c$, (1) can be replaced by a slightly weaker assumption. Assume that for every continuous function $\psi: R^+ \to W$, with range in a compact set $K \subset W$, one has

$$\int_t^{t+1} |h(\psi(s), s)|^p \, ds \to 0 \quad \text{as } t \to \infty. \tag{4}$$

6. Show that $h_t \to 0$ as $t \to \infty$ in $\mathcal{T}_c$.

7. Show that if for every continuous function $\psi: R^+ \to W$, with range in a compact set $K \subset W$, one has

$$\int_0^\infty |h(\psi(s), s)|^p \, ds < \infty, \tag{5}$$

then $h$ satisfies (4).

VIII.C. **Stability.** In this section we shall show that solutions of the limiting equations inherit the stability properties of solutions of the given equation. We shall also show that to a limited extent, the converse is true.

The following lemma will play a central role in this analysis. We assume here that the origin $O$ lies in $W$.

VIII.6 LEMMA. (A) *Let $f$ be a regular function in $C(W \times R, R^n)$ with $f(0,t) = 0$ for all $t \geq 0$. Then for every $g$ in $\Omega_f$ one has $g(0,t) = 0$ for all $t$.*

(B) *Let $\alpha(r,t)$ be a nonnegative function defined on $\{(r,t):$ $0 \leq r \leq a,\ 0 \leq t < \infty\}$, which is nondecreasing in $r$. Let $f$ be a regular function in $C(W \times R, R^n)$ and assume that*

$$|\phi(x, f_\tau, t)| \leq \alpha(|x|, t), \qquad (|x| \leq a,\ 0 \leq \tau,\ 0 \leq t). \tag{6}$$

*Then for every $g$ in $\Omega_f$ one has*

$$|\phi(x, g, t)| \leq \alpha(|x|, t), \qquad (|x| \leq a,\ 0 \leq t). \tag{7}$$

*Proof:* Statement (A) is obvious. To prove (B) we will let $\varepsilon > 0$ be given and let $I$ be a compact interval in $R^+$. For every $g$ in $\Omega_f$ we can find a neighborhood $U$ – in $C(W \times R, R^n)$ – such that for any admissible $h$ in $U$ one has

$$|\phi(x, g, t) - \phi(x, h, t)| \leq \varepsilon\ , \qquad (|x| \leq a,\ t \in I). \tag{8}$$

(This is a consequence of Kamke's Lemma, Lemma IV.2.) By the definition of $\Omega_f$ we can find a translate $f_\tau$ such that $f_\tau \in U$. Then by combining (6) and (8) one has

$$|\phi(x, g, t)| \leq \varepsilon + \alpha(|x|, t)\ , \qquad (|x| \leq a,\ t \in I)..$$

Now let $\varepsilon \to 0$ and we get

$$|\phi(x, g, t)| \leq \alpha(|x|, t), \qquad (|x| \leq a, \ t \in I).$$

Since the right side of the last inequality does not depend on $I$, we have proved (7). Q.E.D.

Let $f$ be an admissible function in $C(W \times R, R^n)$ and assume that $W$ contains the origin $O$ and that $f(0,t) = 0$, $t \geq 0$. We shall say that the null solution $\phi(0, f, t) = 0$ is

(i) **stable** if

$$|\phi(x, f, t)| \leq \alpha(|x|), \qquad (|x| \leq a, \ 0 \leq t),$$

where $\alpha(r)$ is a nonnegative, continuous, nondecreasing function defined on $0 \leq r \leq a$ with $\alpha(0) = 0$;

(ii) **uniformly stable** if

$$(9) \qquad |\phi(x, f_\tau, t)| \leq \alpha(|x|), \qquad (|x| \leq a, \ 0 \leq \tau, \ 0 \leq t),$$

where $\alpha$ is as in (i);

(iii) **asymptotically stable** if it is stable and $\phi(x, f_\tau, t) \to 0$ as $t \to \infty$ whenever $|x| \leq a$ and $0 \leq \tau$;

(iv) **uniformly asymptotically stable** if

$$(10) \qquad |\phi(x, f_\tau, t)| \leq \alpha(|x|)\sigma(t), \qquad (|x| \leq a, \ 0 \leq \tau, \ 0 \leq t),$$

where $\alpha$ is as in (i) and $\sigma(t)$ is a positive, continuous, nonincreasing function defined for $0 \leq t$ with $\sigma(t) \to 0$ as $t \to \infty$;

(v) **exponentially stable** if (10) holds where $\sigma(t) = \exp(-kt)$ for some $k > 0$.

We get the next result as an immediate consequence of Lemma 1.6.

1.7 THEOREM. *Let $f$ be a regular function in $C(W \times R, R^n)$, contains the origin $O$, and assume that $f(0,t) = 0$, $t \geq 0$.*

*If the null solution of* $x' = f(x, t)$ *is uniformly stable* {*uniformly asymptotically stable, exponentially stable*}, *then the null solution of every limiting equation* $x' = g(x, t)$, $g \in \Omega_f$, *is uniformly stable* {*uniformly asymptotically stable, exponentially stable*}.

The concept of stability of an arbitrary solution can be defined by a standard technique. That is, we say that a solution $\phi(t)$ of $x' = f(x, t)$ is ... stable if the null solution of

$$(11) \qquad x' = F(x, t) = f(x + \phi(t), t) - f(\phi(t), t)$$

is ... stable. There is a technical problem which arises here. This occurs when the domain of definition of $F$ does not contain a region of the form $W \times R^+$, where $W$ is a neighborhood of the origin. However, if the solution $\phi(t)$ is positively compact, this difficulty does not occur.

The last theorem has the following generalization.

VIII.8 THEOREM. *Let* $f$ *be a regular function in* $C(W \times R, R^n)$. *If there is a positively compact solution* $\phi(x, f, t)$ *of* $x' = f(x, t)$ *that is uniformly stable* {*uniformly asymptotically stable, exponentially stable*} *then every limiting equation* $x' = g(x, t)$ $(g \in \Omega_f)$ *has a compact solution* $\phi(y, g, t)$ *that is uniformly stable* {*uniformly asymptotically stable, exponentially stable*}.

*Proof:* If $g \in \Omega_f$, then we can find a sequence $\{t_n\}$ in $T$ with $t_n \to \infty$ and $f_{t_n} \to g$, cf. Proposition II.6. By extracting a subsequence, if necessary, we can assume that $\phi(x, f, t_n)$ converges in $W$, say that $\phi(x, f, t_n) \to y$. It follows from Theorem IV.8, that $\phi(x, f, t_n + t) \to \phi(y, g, t)$ uniformly for $t$ in compact subsets of $I_{(y,g)}$, the maximal interval of definition of $\phi(y, g, t)$. Since $\phi(x, f, t)$ remains in a compact set $K \subset W$ for all $t \geq 0$, it is easy to see that $\phi(y, g, t) \in K$ for all $t$, that is $\phi(y, g, t)$ is a compact solution of the limit equation $x' = g(x, t)$. It remains to show that $\phi(y, g, t)$ has the appropriate stability property.

Let $F(x, t)$ be given by (11) where $\phi(t) = \phi(x, f, t)$. It follows from our choice of the sequence $\{t_n\}$ that $F_{t_n}$ converges and

$$F_{t_n}(x, t) \to g(x + \phi(y, g, t), t) - g(\phi(y, g, t), t) = G(x, t) \ .$$

In other words, the solution $\phi(y, g, t)$ of $x' = g(x, t)$ is ... stable if and only if the null solution of $x' = G(x, t)$ is ... stable. The result now follows from Theorem VIII.7. Q.E.D.

The last theorem illustrates how a property of solutions of a given equation $x' = f(x, t)$ is inherited by some solutions of the limiting equations. In the special case where $f \in \Omega_f$, this also leads to further information about the solutions of the original equation. Reversing the role of the given equation and the limiting equations is a bit harder. Consider the following example:

$$x' = \begin{cases} [t \log t]^{-1} x \ , & 2 \le t \\ [2 \log 2]^{-1} x \ , & t < 2 \ . \end{cases} \tag{12}$$

The solutions of (12) are neither positively compact, nor stable. However, the solutions of the limiting equation (there is only one) $x' = 0$ are compact and uniformly stable.

For asymptotic stability more can be said. We formulate the next result for the case of the null solution of $x' = f(x, t)$. It can be extended, in the spirit of Theorem VIII.8 to the case of an arbitrary positively compact solution of $x' = f(x, t)$.

VIII.9 THEOREM. *Let $f$ be a regular function in $C(W \times R, R^n)$ where the motion $f_t$ is positively compact. Assume that $W$ contains* *the origin* 0 *and that* $f(0, t) = 0$ *for* $t \geq 0$. *Assume further that*

(i) *the null solution of* $x' = f(x, t)$ *is uniformly stable, and*

(ii) *the null solution of every limiting equationsis asymptotically stable and there is a constant* $b > 0$ *such that* $|\phi(x, g, t)| \to 0$ *as* $t \to \infty$ *whenever* $|x| \leq b$ *and* $g \in \Omega_f$. *Then the null solution of the given equation* $x' = f(x, t)$ *is asymptotically stable.*

*Proof:* Let $\alpha(r)$ be a nonnegative, continuous, nondecreasing function defined for $0 \leq r \leq a$ with $\alpha(0) = 0$ and satisfying

$$|\phi(x, f_\tau, t)| \leq \alpha(|x|), \qquad (|x| \leq a,\ 0 \leq \tau,\ 0 \leq t).$$

Such an $\alpha$ exists because the null solution of $x' = f(x, t)$ is uniformly stable. Then by Lemma VIII.6 one has

$$(13) \qquad |\phi(x, g, t)| \leq \alpha(|x|), \qquad (|x| \leq a,\ 0 \leq t),$$

for every $g \in \Omega_f$.

Now restrict the flow $\pi$ on $W \times H(f)$ to the invariant subspace $W \times \Omega_f$ and let $M = \{0\} \times \Omega_f$. $M$ is a compact invariant set in the flow $\pi$ and we claim that it is a stable attractor in the flow $\pi$ on $W \times \Omega_f$.

In order to show that $M$ is stable we refer to (13). For every $\varepsilon > 0$ we let $U_\varepsilon$ be the neighborhood of $M$ consisting of all $(x, g)$ in $W \times \Omega_f$ with $|x| < \varepsilon$. Now choose $\delta = \delta(\varepsilon) > 0$ so that $\alpha(r) < \varepsilon$ whenever $0 \leq r < \delta$. Then (13) implies that

$$\pi(U_\delta, t) \subset U_\varepsilon$$

for all $t \geq 0$, so $M$ is stable.

In order to show that $M$ is an attractor we will use hypothesis (ii). If $(x, g) \in U_b$, where $b$ is given in hypothesis (ii), then

$$\pi(x, g, t) = (\phi(x, g, t), g_t) \to \{0\} \times \Omega_f = M$$

as $t \to \infty$. Let $A(M)$ denote the region of attraction of $M$, that is,

$$A(M) = \{(x, g) \in W \times \Omega_f : \pi(x, g, t) \to M \text{ as } t \to \infty\}.$$

Since the open neighborhood $U_b$ lies in $A(M)$, we see that $M$ is an attractor.

Now consider a solution $\phi(x, f, t)$ of $x' = f(x, t)$, where $|x|$ satisfies $|x| \leq a$ and $\alpha(|x|) \leq b$. Since $|x| \leq a$ the solution $\phi(x, f, t)$ is positively compact, and the $\omega$-limit set $\Omega = \Omega_{(x,f)}$ is a nonempty, compact, invariant set in $W \times \Omega_f$. Also, since $|\phi(x, f, t)| \leq \alpha(|x|) \leq b$, the set $\Omega$ lies in $A(M)$. It follows from Lemma II.12 that $\Omega \subset M$. Therefore the positive limit set satisfies $L^+_{(x,f)} = P(\Omega) = \{0\}$, that is $\phi(x, f, t) \to 0$ as $t \to \infty$. Q.E.D.

Hypothesis (ii) deserves a comment. If one assumes that the null solution of every limiting equation is asymptotically stable, this implies that for every $g$ in $\Omega_f$, there is a $b$ (depending on $g$) such that $|\phi(x, g, t)| \to 0$ as $t \to \infty$ whenever $|x| \leq b$. Hypothesis (ii) requires that the constant $b$ can be chosen independent of $g$.

The last theorem has several shortcomings, perhaps the most important one being that we have to assume that the null solution of the given equation is uniformly stable. The author feels that the following statement is true but is unable to prove it.

"Let $f$ be a regular function in $C(W \times R, R^n)$ where the motion $f_t$ is positively compact. Assume that $W$ contains the origin 0 and that $f(0, t) = 0$ for $t \geq 0$. Assume further that there exist continuous positive, monotone functions $\alpha(r)$ and $\sigma(t)$ defined on $0 \leq r \leq a$, $0 \leq t < \infty$, respectively, with $\alpha(r) \to 0$ as $r \to 0$ and $\sigma(t) \to 0$ as $t \to \infty$, and such that one has

$$|\phi(x, g, t)| \leq \alpha(|x|)\sigma(t), \qquad (|x| \leq a,\ 0 \leq t),$$

for every $g$ in $\Omega_f$. Then the null solution of $x' = f(x, t)$ is uniformly asymptotically stable.''

The following result, which is a special case of the above conjecture, is a generalization of a theorem of L. Markus [18].

VIII.10 THEOREM. *Let $f$ be a regular function in $C(W \times R, R^n)$ that is asymptotically autonomous. Assume that $W$ contains the origin 0 and that $f(0, t) = 0$ for $t \geq 0$. Assume further that the null solution of the limiting equation $x' = g(x)$ is uniformly asymptotically stable. Then the null solution of $x' = f(x, t)$ is uniformly asymptotically stable.*

*Proof:* (Outline) Since $f$ is asymptotically autonomous, we can wirte it as $f(x, t) = g(x) + h(x, t)$, where $h(x, t) \to 0$ as $t \to \infty$, uniformly on compact sets in $W$, cf. Theorem VIII.3. Since the null solution of the limiting equation $x' = g(x)$ is uniformly asymptotically stable, there exists (cf. T. Yoshizawa [33, p. 100]) a Lyapunov function $V(x)$ such that $V$ is $C^1$ and

$$\alpha(|x|) \leq V(x) \leq \beta(|x|), \tag{14}$$

where $\alpha$ and $\beta$ are continuous, strictly increasing functions with $\alpha(0) = \beta(0) = 0$. Furthermore $V$ satisfies

$$\frac{\partial V}{\partial x} \cdot g(x) \leq -cV(x)$$

where $c > 0$, and $|\partial V/\partial x| \leq K$ when $|x| \leq a$. Because of the property on $h(x, t)$ there exists positive, continuous, monotone functions $\gamma(r)$ and $\sigma(t)$ with $\gamma(r) \to 0$ as $r \to 0$, $\sigma(t) \to 0$ as $t \to \infty$ and

$$|h(x, t)| \leq \gamma(|x|)\sigma(t), \qquad (|x| \leq a, \ 0 \leq t).$$

(Cf. Exercise 4 below.) If $\phi(t)$ is a solution of $x' = f(x, t)$ with $|\phi(t)| \leq a$ and $V(t) = V(\phi(t))$, then

$$V'(t) = \frac{\partial V}{\partial x} \cdot f(x, t)\Big|_{x = \phi(t)} \leq -cV(t) + K\gamma(a)\sigma(t) .$$

This differential inequality implies that

$$V(t) \leq e^{-ct}\left[V(0) + K\gamma(a) \int_0^t e^{cs}\sigma(s)ds\right] \tag{15}$$

We leave it as an exercise to show that the uniform asymptotic stability of the null solution of $x' = f(x, t)$ follows from (14) and (15). Q.E.D.

We introduce here another type of stability which will be used in the next section. This stability arises in perturbation problems where one wants to compare the solution of the perturbed equation

$$x' = f(x, t) + h(x, t) = g(x, t)$$

with those of the unperturbed equation

$$x' = f(x, t) .$$

In these problems, the function $h$ is usually small in some sense. Usually one is not interested so much in a single perturbed equation but rather a class $\mathcal{G}$ of perturbed equations. Assume now that $f$ is an admissible function in $C(W \times R, R^n)$ and $\mathcal{G}$ denotes a set of admissible functions $g$ from $C(W \times R, R^n)$. We shall say that a solution $\phi(x, f, t)$ of $x' = f(x, t)$ is **stable with respect to** $\mathcal{G}$ if there is a continuous real-valued function $\alpha(r, s)$ defined for $0 \leq r < a_1$, $0 \leq s < a_2$, such that $\alpha$ is increasing in each variable separately, $\alpha(0,0) = 0$ and

(16) $$|\phi(x,f,\tau+t) - \phi(y,g,t)| \leq \alpha(|\phi(x,f,\tau) - y|, \rho(f_\tau, g))$$

for all $t \geq 0$, $\tau \geq 0$ and $g \in \mathcal{G}$. ($\rho$ denotes a metric on $C(W \times R, R^n)$ that generates the topology of uniform convergence on compact sets.) This stability means that if $y$ is "close to $\phi(x,f,\tau)$" and $g$ is close to $f_\tau$, then the solutions $\phi(x,f,\tau+t)$ and $\phi(y,g,t)$ remain close for all $t \geq 0$.

If the solution $\phi(x,f,t)$ of $x' = f(x,t)$ is stable with respect to $\mathcal{G}$, then for every $\tau \geq 0$ the solution

$$\phi(\phi(x,f,\tau), f_\tau, t) = \phi(x,f,\tau+t)$$

of $x' = f_\tau(x,t)$ is stable with respect to $\mathcal{G}$. Moreover, we can prove the following result:

VIII.11 THEOREM. *Let $f \in C(W \times R, R^n)$ be a regular function and let $\phi(x,f,t)$ be a positively compact solution of $x' = f(x,t)$ that is stable with respect to a set $\mathcal{G}$. If $(x^*, f^*) \in \Omega_{(x,f)}$, then the compact solution $\phi(x^*, f^*, t)$ of the limiting equation $x' = f^*(x,t)$ is stable with respect to $\mathcal{G}$. Moreover, if $\alpha(r,s)$ satisfies (16) for $\phi(x,f,t)$, then the same $\alpha(r,s)$ can be used for $\phi(x^*, f^*, t)$.*

*Proof:* By using the change of variables in (11) we can assume that $\phi(x,f,t) = 0$ for all $t \geq 0$. In this case, (16) becomes

$$|\phi(y,g,t)| \leq \alpha(|y|, \rho(f_\tau, g))$$

for all $t \geq 0$ and all $g \in \mathcal{G}$. We will now show that the null solution of every limiting equation is stable with respect to $\mathcal{G}$. Indeed, if $f^* \in \Omega_f$, then there is a sequence $\{f_{\tau_n}\}$ with $\tau_n \to \infty$ and $f_{\tau_n} \to f^*$. By continuity we get $\rho(f_{\tau_n}, g) \to \rho(f^*, g)$ and

$$\alpha(|y|, \rho(f_{\tau_n}, g)) \to \alpha(|y|, \rho(f^*, g)).$$

Therefore

$$|\phi(y, g, t)| \leq \alpha(|y|, \rho(f^*, g))$$

for all $t \geq 0$ and every $g \in \mathcal{G}$ and $f^* \in \Omega_f$, which completes the proof. Q.E.D.

VIII.D. **Existence of Periodic and Almost Periodic Solutions.** In this section we shall apply the results of Chapter VII to the problem of determining the existence of periodic and almost periodic solutions of a differential equation

$$x' = f(x, t) \quad . \tag{17}$$

Our results will be formulated primarily in terms of the limiting equations of (17). In the special case where $f \in \Omega_f$, we are able to conclude something about the given equation (17) itself. This special case includes those differential equations that are periodic or almost periodic in $t$.

The first result is a direct application of Theorem VII.2 and Lemma VIII.1.

VIII.12 THEOREM. *Let $f$ be a regular function in $C(W \times R, R^n)$ and assume that some motion $\pi(x, f, t)$ in $W \times H(f)$ is positively compact and uniformly positively Lyapunov stable with respect to $\Omega_{(x,f)}$. Then $f$ is asymptotically almost periodic and the $\omega$-limit set $\Omega_{(x,f)}$ is an a.p. minimal set. In particular, every limiting equation has an almost periodic solution. If, in addition, $f \in \Omega_f$, the equation $x' = f(x, t)$ itself has an almost periodic solution.*

The real problem facing us is to find conditions which are testable and which imply that some motion $\pi(x, f, t)$ is stable in the sense referred to above. In this regard we are able to prove the following result, which gives a sufficient condition for the existence of almost periodic solutions.

VIII.13 THEOREM. *Let* $f$ *be a regular function in* $C(W \times R, R^n)$ *that is asymptotically almost periodic. Let* $\phi(x, f, t)$ *be a positively compact solution of* $x' = f(x, t)$ *that is stable with respect to* $\Omega_f$. *Then every motion* $\pi(y, g, t)$ *in the* $\omega$*-limit set* $\Omega_{(x,f)}$ *is compact and uniformly positively Lyapunov-stable with respect to* $\Omega_{(x,f)}$. *In particular,* $\Omega_{(x,f)}$ *contains an a.p. minimal set. Furthermore every limiting equation of* $x' = f(x, t)$ *has an almost periodic solution. If, in addition* $f$ *is almost periodic in* $t$, *then* $x' = f(x, t)$ *has an almost periodic solution.*

*Proof:* We choose $\alpha(r, s)$ so that (16) is satisfied. From Theorem VIII.11 it follows that the same $\alpha(r, s)$ "works" for every $g$ in $\Omega_f$, that is, one has

$$|\phi(x^*, f^*, t) - \phi(y, g, t)| \leq \alpha(|x^* - y|, \rho(f^*, g)) \qquad (t \geq 0) \tag{18}$$

for every $f^* \epsilon \Omega_f$ where $(x^*, f^*) \epsilon \Omega_{(x,f)}$. There is no restriction on $y$ in (18), so we note that (18) holds whenever $(y, g)$ is in $\Omega_{(x,f)}$. Let $\varepsilon > 0$ be given. Since $\Omega_f$ is an a.p. minimal set, it is equicontinuous (Theorem VI.11), and there is a $\delta > 0$ such that $\rho(f_t^*, g_t) \leq \varepsilon$ whenever $f^*$ and $g$ are in $\Omega_f$ and $\rho(f^*, g) \leq \delta$. Since $\alpha(r, s)$ is continuous, $\delta$ can be chosen so that (18) implies that

$$|\phi(x^*, f^*, t) - \phi(y, g, t)| \leq \varepsilon \qquad (t \geq 0)$$

whenever $(x^*, f^*)$ and $(y, g)$ are in $\Omega_{(x,f)}$ and

$$|x^* - y| + \rho(f^*, g) \leq \delta . \tag{19}$$

It follows that (19) implies that

$$|\phi(x^*, f^*, t) - \phi(y, g, t)| + \rho(f_t^*, g_t) \leq 2\varepsilon, \qquad (t \geq 0).$$

In other words, the motion $\pi(x^*, f^*, t)$ is uniformly positively Lyapunov-stable with respect to $\Omega_{(x, f)}$. Since the compact invariant set $\Omega_{(x, f)}$ contains a minimal set we see that this minimal set is a.p. minimal. The remainder of the theorem is now obvious. Q.E.D.

If the function $f(x, t)$ in the last theorem had the form $f = g + h$, where $g_t$ is almost periodic in $C(W \times R, R^n)$ and $h(x, t) \to 0$ as $t \to \infty$, uniformly on compact sets in $W$, then the corresponding motion $\pi(x, f, t)$ would be uniformly positively Lyapunov-stable with respect to $\Omega_{(x, f)}$. Our hypotheses are weaker, so Theorem VIII.13 is an improvement over Theorem VIII.12.

Let us consider now periodic or asymptotically periodic differential equations. The first result gives sufficient conditions for the existence of an almost periodic solution.

VIII.14 THEOREM. *Let $f$ be an admissible function in $C(W \times R, R^n)$, that is $T$-periodic in $t$, and let $\phi(x, f, t)$ be a positively compact solution that is uniformly stable. Then the corresponding motion $\pi(x, f, t)$ in $W \times H(f) = W \times \gamma(f)$ is positively compact and uniformly positively Lyapunov-stable. Also, the equation $x' = f(x, t)$ has an almost periodic solution with range in the positive limit set $L^+_{(x, f)}$ and, furthermore, this solution is uniformly stable.*

*Proof:* Since both the motion $f_t$ in $C(W \times R, R^n)$ and the solution $\phi(x, f, t)$ are positively compact, it follows (cf. Lemma IV.4) that the corresponding motion $\pi(x, f, t)$ is positively compact in

$W \times H(f)$. The uniform stability of $\phi(x, f, t)$ is equivalent to saying that for every $\varepsilon > 0$ there is a $\delta$, $0 < \delta \leq \varepsilon$, such that

(20) $$|\phi(x, f, \tau + t) - \phi(y, f_\tau, t)| \leq \varepsilon \qquad (\tau \geq 0, \quad t \geq 0)$$

provided $|\phi(x, f, \tau) - y| \leq \delta$. Furthermore, since $f_t$ is periodic, $\delta$ can be chosen so that

(21) $$\rho(f_{\tau+t}, f_{\sigma+t}) \leq \varepsilon, \qquad (t \in R)$$

whenever $\rho(f_\tau, f_\sigma) \leq \delta$. Since $\phi$ is continuous one can find an $\eta$, $0 < \eta \leq \delta$, so that $|\tau - \sigma|$ is small and

(22) $$|\phi(x, f, \tau) - \phi(y, f_\sigma, \tau - \sigma)| \leq \delta$$

whenever $\rho(f_\tau, f_\sigma) \leq \eta$ and $|\phi(x, f, \tau) - y| \leq \eta$. With $\varepsilon$, $\delta$ and $\eta$ given as above, choose $f_\sigma \in H(f)$ and $y \in W$ so that $\rho(f_\tau, f_\sigma) \leq \eta$ and $|\phi(x, f, \tau) - y| \leq \eta$. If we set $y^* = \phi(y, f_\sigma, \tau - \sigma)$, then

$$\phi(y^*, f_\tau, t) = \phi(y, f_\sigma, t + \tau - \sigma) .$$

Inequalities (20) and (22) then imply that

$$|\phi(x, f, \tau + t) - \phi(y^*, f_\tau, t)| \leq \varepsilon,$$

which together with (21) implies that $\pi(x, f, t)$ is uniformly positively Lyapunov-stable. The rest of this theorem now follows from Theorems VIII.8 and VIII.12. Q.E.D.

An immediate consequence of the last theorem and Theorem VIII.8 is the following:

VIII.15 COROLLARY. *Let $f$ be a regular function in $C(W \times R, R^n)$ that is asymptotically periodic. If there is a solution $\phi(x, f, t)$ of $x' = f(x, t)$ that is positively compact and uniformly* *stable, then every limiting equation has an almost periodic solution.*

Let us now turn to the question of periodic solutions. Based on the results of Chapter VII, one expects a stronger type of stability to be used here. Our main tool will be Theorem VII.4 which gives a sufficient condition that an $\omega$-limit set be periodic minimal. Let us first determine the meaning of the asymptotic stability in this setting.

VIII.16 LEMMA. *Let $f$ be a regular function in $C(W \times R, R^n)$ and assume that there is a motion $\pi(x, f, t)$ in $W \times H(f)$ that is positively compact and asymptotically stable with respect to $\Omega_{(x,f)}$. Then $f$ is asymptotically periodic.*

*Proof:* It follows from Theorem VII.4 that $\Omega_{(x,f)}$ is a periodic minimal set. Since the projection $Q: W \times H(f) \to H(f)$ maps $\Omega_{(x,f)}$ onto $\Omega_f$. which itself is a periodic minimal set, it follows that $f$ is asymptotically periodic. Q.E.D.

VIII.17 THEOREM. *Let $f$ be an admissible function in $C(W \times R, R^n)$ that is $T$-periodic in $t$. Assume that there is a positively compact solution $\phi(x, f, t)$ that is uniformly asymptotically stable. Then $x' = f(x, t)$ has a $kT$-periodic solution (where $k$ is a positive integer) that is uniformly asymptotically stable.*

*Proof:* The uniform asymptotic stability of $\phi(x, f, t)$ implies that the corresponding motion $\pi(x, f, t)$ in $W \times H(f)$ is asymptotically stable with respect to $W \times H(f) \supset \Omega_{(x,f)}$. Therefore $\Omega_{(x,f)}$ is periodic minimal by Theorem VII.4. If $(y, f) \in \Omega_{(x,f)}$ it follows from Lemma VIII.1 that the solution $\phi(y, f, t)$ is periodic, that is

$kT$-periodic in $t$, where $k$ is some positive integer. The stability of $\phi(y, f, t)$ follows from Theorem VIII.8. Q.E.D.

The next corollary is an immediate consequence of the last result and Theorem VIII.8.

VIII.18 COROLLARY. *Let $f$ be a regular function in $C(W \times R, R^n)$ that is asymptotically periodic and assume that there is a positively compact solution $\phi(x, f, t)$ that is uniformly asymptotically stable. Then every limiting equation has a periodic solution (with a commensurable period) that is uniformly asymptotically stable.*

VIII.E **Behavior Near an Equilibrium Point.** In this section we shall examine the behavior of solutions in the neighborhood of an equilibrium point. We can normalize and assume that the equilibrium point is located at the origin $x = 0$. The solution $\phi(0, f, t) \equiv 0$ is called the **trivial** solution and all other solutions are called **nontrivial.**

VIII.19 THEOREM. *Let $f$ be a regular function in $C(W \times R, R^n)$ and assume that the motion $f_t$ is compact. Assume that $f(0,t) = 0$ for all $t$. Then, for every neighborhood $U$ of the origin $O$, one of the following hold:*

(i) *There is a nontrivial solution $\phi(t)$ of $x' = f(x, t)$ with $\phi(t) \to 0$ as $t \to \infty$, or*

(ii) *for some equation $f^*$ in the hull $H(f)$ there is a nontrivial solution $\phi^*(t)$ of $x' = f^*(x, t)$ with $\phi^*(t) \to 0$ as $t \to -\infty$, or*

(iii) *for some equation $f^*$ in the hull $H(f)$ there is a nontrivial solution $\phi^*(t)$ of $x' = f^*(x, t)$ with $\phi^*(t) \in U$ for all $t$ in $R$.*

*Proof:* Let $U$ be any neighborhood of the origin and let $V$ be an open sphere of radius $r$, centered at $0$, with closure $\overline{V} \subset U$.

Step 1. If there is a nontrivial solution $\phi(t)$ of $x' = f(x, t)$ with $\phi(t) \in \overline{V}$ for all $t \geq 0$, then the positive limit set $L^+(\phi)$ lies in $\overline{V}$. In this case, either $L^+(\phi) = \{0\}$ or $L^+(\phi) \neq \{0\}$. If $L^+(\phi) = \{0\}$, this is conclusion (i). If $L^+(\phi) \neq \{0\}$, there is a $y \in L^+(\phi)$, $y \neq 0$. Conclusion (iii) then follows from Theorem IV.6.

Step 2. Assume that every nontrivial solution $\phi(t)$ starting in $V$ at $t = 0$ leaves $\overline{V}$ for some $t > 0$. This means that for $x \in V$, $x \neq 0$, there is a unique $t_x > 0$ such that

$$\phi(x, f, t) \in V, \ (0 \leq t < t_x), \text{ and } \phi(x, f, t_x) \in \partial V \tag{23}$$

where $\partial V$ denotes the boundary of $V$. Now choose a sequence $\{x_n\} \subset V$ with $x_n \to 0$ and let $t_n = t_{x_n}$ denote the corresponding times satisfying (23). Let $y_n = \phi(x_n, f, t_n)$, then $|y_n| = r > 0$. We claim that $t_n \to +\infty$. Indeed if some subsequence (which we shall denote by $\{t_n\}$) is bounded, then there is a $T > 0$ with $0 \leq t_n \leq T$ for all $n$. By the continuity of solutions (Lemma IV.2) one gets $\phi(x_n, f, t) \to \phi(0, f, t) = 0$ uniformly for $0 \leq t \leq T$. But this gives the contradiction

$$0 < r = |y_n| \to 0 \ .$$

Also note that

$$\phi(y_n, f_{t_n}, t) \in \overline{V}, \qquad -t_n \leq t \leq 0. \tag{24}$$

Now choose subsequences so that

$$y_n \to y \in \partial V \quad \text{and} \quad f_{t_n} \to f^* \in H(f) \ .$$

Then by Lemma IV.2 one has

$$\phi(y_n, f_{t_n}, t) \to \phi(y, f^*, t)$$

uniformly on compact sets in $R$. It now follows from (24) that $\phi(y, f^*, t) \in \bar{\bar{V}}$ for all $t \leq 0$. Hence by Step 1, where we replace $t$ with $-t$, we see that either (ii) or (iii) hold. Q.E.D.

It is evident that one can interchange the roles of $t$ and $-t$ in the conclusions of the last result. For further discussion of ramifications of this result see Exercise 8 and [31].

*Exercises 1.* Prove the following analogue of Lemma VIII.2: Show that if $f_t$ is a compact recurrent motion in $H(f)$ and $\phi(x, f, t)$ is a compact recurrent solution, then the motion $\pi(x, f, t)$ is compact and recurrent. (Hint: Derive the analogue of Theorem V.9 for recurrent functions.)

*2.* Is an analogue of Lemma VIII.2 valid for Poisson-stable solutions?

*3.* Let $f \in C(W \times R, R^n)$ and assume that the motion $f_t$ is positively compact. Show that $f$ is asymptotically Poisson-stable if and only if there is a Poisson-stable point $g$ in $\Omega_f$ with $H(g) = \Omega_f$.

*4.* Let $h \in C(R^n \times R, R^n)$ where $h(x, t) \to 0$ as $t \to \infty$ uniformly on compact sets in $R^n$ and assume that $h(0,t) = 0$. Define $\gamma(r)$ and $\sigma(t)$ to be nonnegative and satisfying

$$\gamma(r)^2 = \sup\{|h(x, t)| : |x| \leq r, \ 0 \leq t\}$$
$$\sigma(t)^2 = \sup\{|h(x, t)| : |x| \leq 1\} .$$

Show that

$$|h(x, t)| \leq \gamma(|x|)\sigma(t) , \quad (|x| \leq 1, \ 0 \leq t) .$$

Show that $\gamma$ and $\sigma$ are continuous and monotone. Show that $\gamma(r) \to 0$ as $r \to 0$, $\sigma(t) \to 0$ as $t \to \infty$.

5. Let $\phi(x, f, t)$ be a positively compact solution of $x' = f(x, t)$ that is uniformly asymptotically stable. Assume that $f = g + h$ where $g_t$ is almost periodic in $C(W \times R, R^n)$ and $h(x, t) \to 0$ as $t \to \infty$, uniformly on compact sets in $W$. Show that $\phi(x, f, t)$ is stable with respect to $\Omega_f$. Show that $x' = g(x, t)$ has an almost periodic solution.

6. Theorem VIII.17 asserts the existence of a $kT$-periodic solution. Give an example showing that the integer $k$ may satisfy $k \geq 2$.

7. Let $f$ be a regular function in $C(W \times R, R^n)$ that is asymptotically autonomous. Assume that there is a positively compact solution $\phi(x, f, t)$ that is uniformly asymptotically stable. Show that $\lim_{t \to \infty} \phi(x, f, t)$ exists, call this limit $y$. Show that the point $y$ is a uniformly asymptotically stable equilibrium point of the limiting equation.

8. Let $\phi(t)$ be a compact solution of $x' = f(x, t)$ where $f$ is a regular function in $C(W \times R, R^n)$ with compact hull $H(f)$. Derive a result analogous to Theorem VIII.19 for the behavior of solutions near $\phi(t)$.

9. Consider the differential equation

$$x' = Ax + R(x, t) + G(t) \tag{25}$$

where $R(0,t) = 0$, $|R(x, t) - R(y, t)| \leq k|x - y|$ and $|G(t)| \leq B$ for all $x$, $y$ in $R^n$ and $t$ in $R$. Assume that the eigenvalues of $A$ all have negative real part. Show that if $k$ is sufficiently small, then (25) has a compact, uniformly asymptotically stable solution.

Show that if, in addition, $R$ and $G$ are almost periodic in $t$, then (25) has an almost periodic solution.

*10.* Consider the differential equation

$$x' = f(x, t) + R(x, t) + G(t) \tag{26}$$

where $R$ and $G$ satisfy the conditions of Exercise 9. Assume that $f$ is a $C^2$-function and that the Jacobian, $J = \det|f_x(x, t)|$ is bounded on sets of the form $K \times R$ where $K$ is compact in $R^n$. Assume that $x' = f(x, t)$ has a positively compact, exponentially stable solution. Show that (26) has a compact, exponentially stable solution provided both $k$ and $B$ are sufficiently small.

*11.* Consider the linear differential equations

$$\text{(27a)} \quad x' = A(t)x, \qquad \text{(27b)} \quad x' = A(t)x + b(t)$$

where $x \in R^n$. Let $Y(t)$ be a fundamental matrix solution of (27a) with $Y(0) = I$. We say that (27a) has an *exponential dichotomy* if there is a projection $P$ on $R^n$ and positive constants $K$ and $a$ such that

$$|Y(\tau+t)PY^{-1}(\tau)| \le Ke^{-at}, \quad |Y(\tau)QY^{-1}(\tau+t)| \le Ke^{-at}$$

for $\tau \ge 0$, $t \ge 0$ where $Q = I - P$. (i) Show that if (27a) has an exponential dichotomy, then every equation in the hull of (27a) has an exponential dichotomy. (ii) Assume that $A(t)$ and $b(t)$ are almost periodic and that (27a) has an exponential dichotomy. Show that every equation in the hull of (27a) has an almost periodic solution.

## CHAPTER IX

## *INVARIANT MEASURES*

In this chapter and the next we shall study those flows that have an invariant measure. We shall see that in such a flow almost every point is Poisson stable. Of particular interest to us will be an interpretation of these (abstract) results in the flow constructed in Chapter IV.

IX.A **An Example from Classical Mechanics.** The equations of the classical $n$-body problem can be expressed as a Hamiltonian system

$$\frac{dp}{dt} = \frac{\partial H}{\partial q}, \quad \frac{dq}{dt} = -\frac{\partial H}{\partial p} \tag{1}$$

where $H$ is an appropriate Hamiltonian function and $p$ and $q$ are vectors of dimension $3n$. If we write (1) as $x' = F(x)$ where $x = (p, q)$ is a vector of dimension $6n$, then it is easy to see that

$$\operatorname{div} F = 0 . \tag{2}$$

The divergence condition (2) holds for any Hamiltonian system, regardless of dimension. The following theorem asserts that any vector field (even nonautonomous vector field *a*) that satisfies (2) is incompressible.

IX.1. THEOREM. (Liouville) *Let $f(x, t)$ be a $C^2$-vector field on $W \times R$ where $W$ is an open set in $R^n$ and assume that*

$$\operatorname{div}_x f = \sum_{i=1}^{n} \partial f_i/\partial x_i = 0 .$$

*for all $(x, t)$ in $W \times R$. Let $D$ be a Lebesgue measurable set in $W$ and let*

$$D_t = \{\phi(x, f, t) : x \in D\}$$

*be the translate of $D$ by solutions of $x' = f(x, t)$. Assume that every solution starting in $D$ can be continued to $D_t$. Then $\mu(D) = \mu(D_t)$, where $\mu$ denotes the Lebesgue measure. (That is, the flow is volume-preserving, or incompressible.)*

*Proof:* It will suffice to prove this when $D$ is a compact subset of $W$. Define

$$I(t) = \mu(D_t) = \int_{D_t} dx$$

where $\int$ denotes the Lebesgue integral. We will now show that $I'(t) \equiv 0$.

Since every solution starting in $D$ can be continued to $D_t$, they can be continued to $D_{t+h}$, where $|h|$ is positive but small. Also

(3) $$I(t + h) = \int_{D_{t+h}} d\bar{x} :$$

We will now perform a change of variables in the region $D_{t+h}$ and write $I(t + h)$ as an integral over $D_t$. Since the mapping along solution curves defines a diffeomorphism of $D_t$ onto $D_{t+h}$ we can write this as

$$\bar{x} = \phi(x, f_t, h) .$$

If $J$ denotes the Jacobian of $\overline{x}$ with respect to x, that is

$$J = \det\left(\frac{\partial \overline{x}_i}{\partial x_j}\right)$$

then (3) becomes

$$I(t+h) = \int_{D_t} J\,dx \ .$$

Now as is known, the functions $\partial \overline{x}_i/\partial x_j$ satisfy the variational equation

$$\frac{d}{dh}\frac{\partial \overline{x}_i}{\partial x_j} = \sum_{k=1}^{n} \frac{\partial f_i}{\partial \overline{x}_k}(\overline{x})\frac{\partial \overline{x}_k}{\partial x_j} \ .$$

Hence

$$\frac{\partial \overline{x}_i}{\partial x_j} = \delta_{ij} + h\sum_{k=1}^{n} \frac{\partial f_i}{\partial \overline{x}_k}(x)\left.\frac{\partial \overline{x}_k}{\partial x_j}\right|_{h=0} + o(h)$$

$$= \delta_{ij} + h\frac{\partial f_i}{\partial x_j}(x) + o(h) \ .$$

It then follows that

$$J = 1 + h\sum_{i=1}^{n} \partial f_i/\partial x_i\,(x) + o(h) \ .$$

Therefore

$$I'(t) = \lim_{h\to 0}\frac{1}{h}[I(t+h) - I(t)] = \lim_{h\to 0}\frac{1}{h}\int_{D_t}[h \operatorname{div} f + o(h)]dx \ .$$

It follows from $\operatorname{div} f = o$, that $I'(t) = 0$. Q.E.D.

IX.B **Invariant Measures and Poisson Stability.** Let $\pi$ be a flow on a uniform space $X$. Let $\mu$ be a real-valued measure on $X$. We shall say that $\mu$ is **invariant** (with respect to $\pi$) if for every measurable set $A \subset X$ one has $\mu(A) = \mu(\pi(A, \tau))$ for all $\tau$ with the property that every motion beginning in $A$ can be continued to $t = \tau$, that is $x \in A$ implies that $\tau \in I_x$, where $I_x$ is the interval of definition of the motion $\pi(x, t)$. In the sequel we will be primarily interested in **positive** measures $\mu$, ($\mu(A) \geq 0$ for every measurable set $A$) that are **nontrivial** ($\mu(A) > 0$ for some measurable set $A$). We shall say that a positive measure $\mu$ is **normalized** if $\mu(X) = 1$.

IX.1 THEOREM. *Let $\pi$ be a global flow on a uniform space $X$, where $X$ has a countable basis of neighborhoods.* Let $\mu$ be a normalized invariant measure on $X$. Then almost every point in $X$ is Poisson-stable, that is if $E$ denotes the set of points in $X$ that are not Poisson-stable, then $\mu(E) = 0$.*

*Proof:* Let $A$ be any measurable set. Define $A_\infty$ and $A_{-\infty}$ by

$$A_\infty = A - \bigcup_{n=1}^{\infty} A \cap \pi(A, n)$$

(4)

$$A_{-\infty} = A - \bigcup_{n=-\infty}^{-1} A \cap \pi(A, n) .$$

We claim that $\mu(A_\infty) = \mu(A_{-\infty}) = 0$. Let us prove this for $A_\infty$. Let $A_\infty^n = \pi(A_\infty, n)$. Then $\mu(A_\infty^n) = \mu(A_\infty)$ for all $n$. Since $A_\infty^n \subset \pi(A, n)$ and $A_\infty \subset A$ it follows that

$$A_\infty \cap A_\infty^n = \emptyset . \qquad n = 1,2,\dots .$$

---

* In this case $X$ is metrizable, cf. [15, p.186].

It follows immediately that $A_\infty^n \cap A_\infty^m = \emptyset$ provided $n < m$. In other words

$$A_\infty^n \cap A_\infty^m = \emptyset$$

when $n \neq m$ and $0 \leq n$, $0 \leq m$. Since $\{A_\infty^n, n = 0,1,\ldots\}$ is a family of disjoint sets of equal measure, it follows from $\mu(X) = 1$ that $\mu(A_\infty^n) = 0$ for all $n = 0,1,\ldots$. The proof that $\mu(A_{-\infty}) = 0$ is similar.

Now let $\{U^{(n)}\}$ be a countable basis of neighborhoods in $X$. Construct the sets $U_\infty^{(n)}$ and $U_{-\infty}^{(n)}$ by (4). Let

$$E^+ = \cup_n U_\infty^{(n)}, \quad E^- = \cup_n U_{-\infty}^{(n)}, \quad E = E^+ \cup E^- .$$

Then $E^+$, $E^-$ and $E$ all have measure zero.

We claim now that every point in $x$ in $X - E^+$ is stable $P^-$. Indeed, let $x \in X - E^+$. Then, for any neighborhood $U^{(k)}$ of $x$ in the given basis, there is a natural number $m$ such that $x \in \pi(U^{(k)}, m)$, or $\pi(x, -m) \in U^{(k)}$. Hence $x$ is stable $P^-$. Similarly every point $x$ in $X - E^-$ is stable $P^+$. It follows then that every point in $X - E$ is Poisson stable. Q.E.D.

The fact that the measure $\mu(X)$ is precisely 1, is, of course, not critical. All we really use is the fact that $\mu(X)$ is finite. If $\mu(X)$ is not finite one can derive results related to those above. Cf. [22, pp. 454-459.]

IX.C **Existence of Invariant Measures.** Let us now turn to the question of the existence of invariant measures. We will now assume that $X$ is a compact metric space and $\pi$ a flow on $X$. ($\pi$ is necessarily a global flow.) We will now show that there is a normalized invariant measure $\mu$ on $X$. In order to prove this we shall make use of some concepts from functional analysis. The first is a representation theorem of Riesz.

Let $C(X)$ denote the Banach space of real-valued continuous functions $f$ defined on $X$, with the norm given by

$$\|f\| = \sup\{|f(x)| : x \in X\} .$$

The proof of the next result can be found in [9, p. 265].

IX.2 THEOREM. (Riesz) *Let* $\ell$ *be a bound linear functional defined on* $C(X)$. *Then there is a measure* $\mu$ *such that*

$$\ell(f) = \int_X f(x)\,\mu(dx) \ .$$

*Moreover, if* $\ell$ *is positive, then* $\mu$ *is positive. If* $\ell(1) = 1$, *then* $\mu(X) = 1$.

Another fact we need is that any sequence $\{\mu_n\}$ of normalized measures contains a weakly convergent subsequence. That is, there is a subsequence $\{\mu_{n_k}\}$ and a normalized measure $\mu$ such that

$$\int_X f(x)\mu_{n_k}(dx) \to \int_X f(x)\mu(dx)$$

for every $f$ in $C(X)$. See [9].

IX.3 THEOREM. *Let* $X$ *be a compact metric space and* $\pi$ *a flow on* $X$. *Then there exists a normalized invariant measure on* $X$.

*Proof:* (Outline) First assume that the group $T$ is the reals $R$. Let $\nu$ be any normalized measure on $X$ and for $\tau > 0$ define $\ell_\tau$ by

(5) $$\ell_\tau(f) = \frac{1}{\tau} \int_0^\tau \int_X f(\pi(x,t))\nu(dx)dt \ .$$

$\ell_\tau$ is then a bounded linear function on $C(X)$ and it is positive. Moreover, $\ell_\tau(1) = 1$. Hence by the Riesz Theorem there is a normalized measure $\mu_\tau$ such that

$$\ell_\tau(f) = \int_X f(x)\mu_\tau(dx) \ .$$

It follows from the weak convergence property that there is a sequence $\tau_n$ such that $\tau_n \to \infty$ and $\mu_{\tau_n} = \mu_n$ converges weakly to normalized measure $\mu^*$. We claim that $\mu^*$ is an invariant measure.

If we can show that for every $f$ in $C(X)$ and $\sigma$ in $T$ one has

(6) $$\int_X f(x)\mu^*(dx) = \int_X f(\pi(x,\sigma))\mu^*(dx)$$

this result follows. That is, the integral $\int_X f(x)\mu^*(dx)$ can be extended to hold for all bounded Borel measurable functions. It follows then that (6) holds even for these functions. If $f$ is then the characteristic function of a Borel measurable set $A$, then $f(\pi(\cdot\,,\sigma))$ is the characteristic function of the set $\pi(A,-\sigma)$. Equation (6) then becomes $\mu^*(A) = \mu^*(\pi(A,-\sigma))$.

In order to prove (6) we note that

(7) $$\frac{1}{\tau_n} \int_0^{\tau_n} \int_X f(\pi(x,t))\nu(dx)dt \to \int_X f(x)\mu^*(dx)$$

and

(8) $$\frac{1}{\tau_n} \int_0^{\tau_n} \int_X f(\pi(x,\sigma+t))\nu(dx)dt \to \int_X f(\pi(x,\sigma))\mu^*(dx) \ .$$

Now by applying Fubini's Theorem we get

$$\left| \frac{1}{\tau_n} \int_0^{\tau_n} \int_X [f(\pi(x,t)) - f(\pi(x,\sigma+t))]\nu(dx)dt \right|$$

$$\leq \int_X \left| \frac{1}{\tau_n} \int_0^{\tau_n} [f(\pi(x,t)) - f(\pi(x,\sigma+t))]\,dt \right| \nu(dx)$$

$$\leq \int_X \frac{1}{\tau_n} \left\{ \left| \int_0^{\sigma} f(\pi(x,t))dt \right| + \left| \int_{\tau_n}^{\tau_n+\sigma} f(\pi(x,t))dt \right| \right\} \nu(dx)$$

$$\leq \int_X \frac{1}{\tau_n} 2\|f\| \, |\sigma| \, \nu(dx) = \frac{2\|f\| \, |\sigma|}{\tau_n} \to 0 \; .$$

Since the limits in (7) and (8) are the same, we have proved (6).

The same methods can be used if the group $T$ is $I$, the integers. In this case the integral $\int_0^{\tau} g(s)ds$ becomes the summation

$$\sum_{i=0}^{n-1} g(i) \; .$$

We shall omit the details. Q.E.D.

IX.D **Examples of Invariant Measures.** We have seen that under the divergence condition (2) the Lebesgue measure is invariant for the flow constructed by the solutions of the autonomous differential equation $x' = F(x)$.

Another important example can be constructed for an a.p. minimal set. This is the Haar measure, cf. [17]. The reader will recall that if $G$ is a compact Abelian topological group, then there is precisely one normalized measure $\mu$ such that for every Borel set $A$ one has $\mu(A) = \mu(A + g)$ for every $g \in G$. This measure is the Haar measure.

Now let $E$ be an a.p. minimal set in a flow $\pi$. It follows from Theorem VI.14 that $E$ is a compact Abelian topological group. The group addition and the flow $\pi$ are related by

$$\pi(y, t) = y + \pi(x, t)$$

where x is the identity of the group. Now let $A$ be any Borel set, then $\pi(A, t) = A + \pi(x, t)$, which implies that

$$\mu(\pi(A, t)) = \mu(A + \pi(x, t)) = \mu(A) \quad .$$

Hence $\mu$ is invariant for the flow $\pi$.

IX.E **Nonautonomous Differential Equations.** In this section we shall study the question of the existence of invariant measures for the flow $\pi(x, f, t)$ constructed on $W \times H(f)$ in Chapter IV. Let $f$ be a regular function in $C(W \times R, R^n)$, We shall say that a measure $\nu$ on $W$ is **invariant (with respect to** $f$) if for every measurable set $A \subset W$ one has $\nu(A) = \nu(\phi(A, f, \tau))$ for all $\tau$ with the property that every solution $\phi(x, f, t)$ with $x \in A$ can be continued to $t = \tau$. We shall say that a measure $\nu$ is **bounded** if the measure of every compact set is finite.

IX.4 LEMMA. *Let* $f$ *be a regular function in* $C(W \times R, R^n)$. *Let* $\nu$ *be a bounded measure on* $W$ *that is invariant with respect to* $f$. *Then* $\nu$ *is invariant with respect to each* $f^*$ *in the hull* $H(f)$.

*Proof:* First we shall show that $\nu$ is invariant with respect to each translate $f_\tau$. Since

$$\phi(x, f_\tau, t) = \phi(\phi(x, f_\tau, -\tau), f, \tau + t) \tag{9}$$

it follows that for any measurable set $A$ in $W$ with $A_\sigma = \phi(A, f, \sigma)$ and $B = \phi(A, f_\tau, -\tau)$ one has

$$\phi(A, f_\tau, \sigma) = \phi(B, f, \tau+\sigma) = B_{\tau+\sigma} \ .$$

We want to show that for any $\tau$ one has

$$\nu(\phi(A, f_\tau, \sigma)) = \nu(A) \tag{10}$$

for all $\sigma$ with the property that every solution $\phi(x, f_\tau, t)$, $x \in A$, can be continued to $t = \sigma$. Since $W$ is locally compact and $\nu$ is countably additive it will suffice to prove (10) when $A$ is compact. Now assume that $A$ is compact and fixed. Then for $|\tau|$ sufficiently small, every solution $\phi(x, f_\tau, t)$, $x \in A$, can be continued to $t = -\tau$. Hence $B$ is compact and $B_\tau = A$. Now assume that every solution $\phi(x, f_\tau, t)$, $x \in A$, can be continued to $t = \sigma$. It then follows that every solution $\phi(x, f, t)$, $x \in B$, can be continued to $t = \tau + \sigma$. Since $\nu$ is invariant with respect to $f$ it follows that $\nu(B) = \nu(B_{\tau+\sigma})$. This holds in particular for $\sigma = 0$, that is

$$\nu(A) = \nu(B_\tau) = \nu(B) = \nu(B_{\tau+\sigma}) \ ,$$

which is (10).

We have shown that (10) holds when $|\tau|$ is small. However, by replacing $f$ with $f_\tau$ and repeating the above argument, one can show that (10) holds when $\tau$ is replaced by $\tau + \tau'$ and $|\tau'|$ is small. It is easy to see that in this manner, we can establish (10) for any $\tau$.

Now let $f^*$ be any function in the hull $H(f)$. We then want to prove (10) where $f_\tau$ is replaced by $f^*$. Again we can assume that $A$ is compact. Since $f^* \in H(f)$ there is a sequence of translates $\{f_{\tau_n}\}$ in $\gamma(f)$ such that $f_{\tau_n} \to f^*$. It follows from Kamke's Lemma (Lemma IV.2) that

(11) $$\phi(x, f_{\tau_n}, t) \to \phi(x, f^*, t) \ .$$

If one assumes that every solution $\phi(x, f^*, t)$, $x \in A$, can be continued to $t = \sigma$, it follows that the convergence in (11) is uniform for $x \in A$ and $t$ in the interval between 0 and $\sigma$. Let

$$B = \phi(A, f^*, \sigma) \text{ and } B_n = \phi(A, f_{\tau_n}, \sigma) \ .$$

Since $\nu$ is invariant with respect to $f_{\tau_n}$ it follows that $\nu(A) = \nu(B_n)$ ($B_n$ is compact for $n$ sufficiently large ). Furthermore, one has $B_n \to B$ in the Hausdorff sense, that is for every $\varepsilon > 0$ there is an $N$ such htat

$$B_n \subset S(B, \varepsilon) \text{ and } B \subset S(B_n, \varepsilon)$$

for $n \geq N$. (Cf. [26].) ($S(B, \varepsilon)$ denotes the $\varepsilon$-neighborhood of $B$.) This implies that

$$\nu(B_n) \leq \nu(S(B, \varepsilon)) \text{ and } \nu(B) \leq \nu(S(B_n, \varepsilon))$$

for $n \geq N$. Since $\nu(A) = \nu(B_n)$ and since $\nu(S(B, \varepsilon)) \to \nu(B)$ as $\varepsilon \to 0$, the first inequality implies that $\nu(A) \leq \nu(B)$. We have thus shown that for every compact set $A$ one has

(12) $$\nu(A) \leq \nu(\phi(A, f^*, \sigma)) \qquad (f^* \in H(f))$$

for any $\sigma$ with the property that every solution $\phi(x, f^*, t)$, $x \in A$, can be continued to $t = \sigma$. Since

$$A = \phi(\phi(A, f^*, \sigma), f^*_\sigma, -\sigma)$$

it follows from (12) that $\nu(\phi(A, f^*, \sigma)) \leq \nu(A)$. Q.E.D.

IX.5 COROLLARY. *Let $f$ be a regular function in $C(W \times R, R^n)$ that is $C^2$ in x. Assume that $\text{div}_x f = 0$ for all $(x, t)$ in $W \times R$. Then the Lebesgue measure on $W$ is invariant with respect to each $f^*$ in the hull $H(f)$.*

*Proof:* This follows directly from Theorem IX.1 and Lemma IX.4. Q.E.D.

It should be noted that this corollary is an extension of Liouville's Theorem since the function $f^*$ may fail to be differentiable so that $\text{div}_x f^*$ is not defined.

The next theorem is an existence theorem for measures on $W \times H(f)$ that are invariant with respect to the flow $\pi(x, f, t)$ defined in Chapter IV.

IX.6 THEOREM. *Let $f$ be a regular function in $C(W \times R, R^n)$ and let $\nu$ be a bounded measure on $W$ that is invariant with respect to $f$. Assume that $f_t$ is almost periodic. Then the product measure $m = \nu \times \mu$, where $\mu$ is the Haar measure on $H(f)$, is invariant for the flow $\pi(x, f, t) = (\phi(x, f, t), f_t)$ on $W \times H(f)$.*

*Proof:* Since the measure $m$ is countably additive and $W \times H(f)$ is locally compact it will suffice to show that for every compact set $A$ in $W \times H(f)$ of the form $A = B \times C$ (where $B \subset W$, $C \subset H(f)$) one has $m(A) = m(\pi(A, \tau))$, provided every motion $\pi(x, g, t)$ $(x \in B, g \in C)$ can be continued to $t = \tau$.

Let $A_\tau = \pi(A, \tau)$ and $C_\tau = \{g_\tau : g \in C\}$. Then

$$A_\tau = \{(\phi(x, g, \tau), g_\tau) : x \in B, g \in C\} .$$

The section of $A_\tau$ found by fixing $g_\tau$ is $\phi(B, g, \tau)$ (See Figure 1 on facing page.) Since $\nu$ is invariant with respect to $g$

(Lemma IX.4), it follows that $\nu(B) = \nu(\phi(B, g, \tau))$. That is, every section $\phi(B, g, \tau)$ of $A_\tau$ has the same $\nu$-measure. By Fubini's

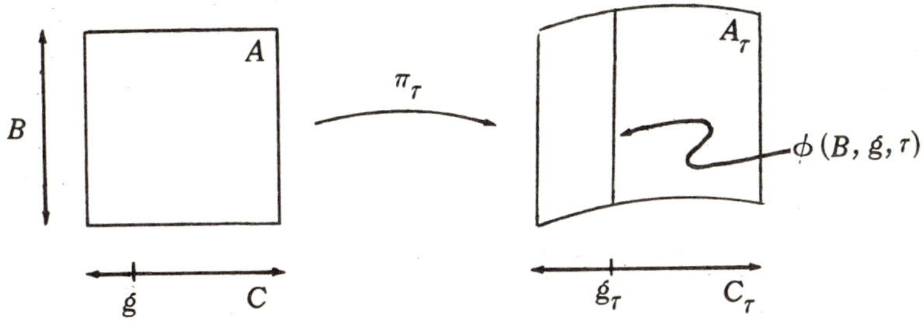


Figure 1.

Theorem one has

$$m(A_\tau) = \int_{A_\tau} 1\, dm = \int_{C_\tau} \left( \int_{\phi(B, g, \tau)} 1\, d\nu \right) d\mu$$

$$= \int_{C_\tau} \nu(B) d\mu = \nu(B)\mu(C_\tau) \ .$$

However, the Haar measure $\mu$ is invariant, that is $\mu(C_\tau) = \mu(C)$. One then has $m(A) = \nu(B)\mu(C) = m(A_\tau)$. Hence $m$ is invariant. Q.E.D.

It is evident that the last theorem has the following generalization.

IX.7 THEOREM. *Let $f$ be a regular function on $C(W \times R, R^n)$ and assume that the motion $f_t$ is compact. Assume that there is a bounded measure $\nu$ on $W$ that is invariant with respect to $f$ and* *let $\mu$ be any invariant measure on $H(f)$. Then the product measure $m = \nu \times \mu$ is invariant for the flow $\pi(x, f, t)$ on $W \times H(f)$.*

Let us now consider a specific application of these results. In this application we shall be interested in differential equations of the form

$$x' = f(x, t) \tag{13}$$

where $f$ is regular in $C(W \times R, R^n)$ and $f_t$ is almost periodic. We will also assume that $f$ is a $C^2$-function in $x$ and $\text{div}_x f = 0$ for all $(x, t)$ in $W \times R$. Equations of this type arise in celestial mechanics when one considers the motion of a satellite (an object of zero gravitational mass) in the gravitational field of $n$-bodies, which themselves are moving in an almost periodic orbit. We ask the following questions: Are any solutions of (13) Poisson stable? How many solutions are Poisson stable? The answer to these questions, as formulated below, we find to be surprising and interesting. We shall assume that the set $W$ is bounded in $R^n$ and that the flow $\pi(x, g, t)$ on $W \times H(f)$ is a global flow, that is, every solution $\phi(x, g, t)$ $(x \in W,\ g \in H(f))$ is defined for all $t$ in $R$.

IX.8 THEOREM. *Let $W$ be a bounded set in $R^n$ and let $f$ be a regular function in $C(W \times R, R^n)$ where the motion $f_t$ is almost periodic. Assume that $f$ is $C^2$ in $x$ and $\text{div}_x f = 0$ for all $(x, t)$ in $W \times R$, and assume that the flow*

$$\pi(x, f, t) = (\phi(x, f, t), f_t)$$

*is a global flow. Let $\nu$ be the Lebesgue on $W$ and $\mu$ the Haar measure on $H(f)$. Then except for a set $E$ in $H(f)$ of (Haar) measure zero, the set*

$$\{x \in W:\ \phi(x, g, t) \text{ is **not** Poisson stable}\}$$

*has (Lebesgue) measure zero. In other words, the probability that an arbitrary solution* $\phi(x, g, t)$ *is not Poisson stable is zero. Furthermore, for every* $g$ *in* $H(f) - E$ *the set*

$$A(g) = \{x \in W: \phi(x, g, t) \text{ is Poisson stable}\} \tag{14}$$

*is everywhere dense in* $W$.

Before we prove this result it may be helpful to reformulate the conclusion. The terminology we use is suggested by the celestial mechanical interpretation. For any $g$ in $H(f)$ we shall refer to the differential equation $x' = g(x, t)$ as a **world.** We shall say that a world $g$ is **good** if the set $A(g)$ is everywhere dense in $W$. Then the theorem says that "except for a set of (Haar) measure zero, every world is good." The question of whether a given world $x' = f(x, t)$ in $H(f)$ is good is not resolved by this result, except in the special case where $f$ is periodic in $t$. (Cf. Corollary IX.9.)

*Proof:* Since $W$ is bounded we can assume that the measures $\nu$ and $\mu$ are normalized, that is $\nu(W) = \mu(H(f)) = 1$. It follows from Theorem IX.6 that the product measure $m = \nu \times \mu$ is invariant for the flow $\pi(x, g, t)$ on $W \times H(f)$. Since $m$ is normalized, it follows from Theorem IX.1 that the set

$$\{(x, g) \in W \times H(f): \pi(x, g, t) \text{ is Poisson stable}\}$$

has $m$-measure 1 in $W \times H(f)$, that is

$$m(\{(x, g): \pi(x, g, t) \text{ is Poisson stable}\}) = m(W \times H(f)) .$$

Define $E$ to be the set of those $g$ in $H(f)$ for which the section

$$B(g) = \{x \in W: \pi(x, g, t) \text{ is Poisson stable}\}$$

has (Lebesgue) measure $< 1$. It follows from Fubini's Theorem that $\mu(E) = 0$. Therefore, for every $g$ in $H(f)-E$, one has

$$\nu(B(g))=1 \text{ and } \nu(W-B(g)) = 0 .$$

It follows from Lemma VIII.1 that $B(g) \subset A(g)$ where $A(g)$ is given by (14). Hence $\nu(A(g)) = 1$ and $\nu(W-A(g)) = 0$. Since the Lebesgue measure of a nonempty open set $U$ in $W$ is positive, $\nu(A(g)) = 1$ implies that $A(g)$ is dense in $W$. Q.E.D.

Many variations of the last result are possible. One can replace the Lebesgue measure on $W$ with any finite positive measure $\nu$ that is invariant with respect to $f$. Also one can replace the assumption that $f_t$ is almost periodic with $f_t$ being compact. In this case $\mu$ would be any invariant measure on $H(f)$. The existence of such a $\mu$ is guaranteed by Theorem IX.3.

If $f$ is periodic in $t$ one can show that the set $E$ in Theorem IX.8 is empty. That is, we can prove the following:

IX.9 COROLLARY. *In addition to the hypotheses of Theorem IX.8, assume that* $f(x,t)$ *is periodic in* $t$. *Then for every* $g$ *in* $H(f)$ $(=\gamma(f))$ *the set*

$$W - A(g)=\{x \in W: \ \phi(x,g,t) \text{ is } \textbf{not} \text{ Poisson stable}\}$$

*has (Lebesgue) measure zero.*

*Proof:* In the proof of Theorem IX.8 one observes that the set $E$ is an invariant set for the flow on $H(f)$. If $f$ is periodic in $t$, the only invariant sets in $H(f) = \gamma(f)$ are $\gamma(f)$ and the empty set. Since $\mu(E) = 0$, it follows that $E = \emptyset$. Q.E.D.

*Exercises 1.* Let $M(x)$ be a $C^1$ real-valued function defined on $W$, an open set in $R^n$. Let $f(x)$ be a $C^2$ vector field on $W$ with div $Mf = 0$. Show that the measure $\mu$ given by $\mu(D) = \int_D M(x)dx$ is invariant under the flow generated by $x' = f(x)$.

2. Let $E$ be an a.p. minimal set in a flow $\pi$ and let $\mu$ be a normalized invariant measure for $\pi$. Show that $\mu$ is the Haar measure on $E$. (Hint: Show that $\mu(A) = \mu(A + y)$ for every Borel set $A \subset E$ and every $y \in E$. Then use the fact that the Haar measure is unique.)

# CHAPTER X

## *ERGODIC THEORY*

In these lectures we shall study only a small corner of the ergodic theory. We shall be particularly interested in two things:
1. The ergodic theorem of Birkhoff for abstract flows and
2. an interpretation of this for the flow $\pi(x, f, t)$ constructed in Chapter IV.

X.A **Birkhoff Ergodic Theorems.** Let $X$ be a uniform space and let $\pi$ be a global flow on $X$. Assume that there is a normalized invariant measure $\mu$ on $X$. For $1 \leq p < \infty$, consider the Lebesgue spaces $L_p(X, \mu)$ of real-valued functions $f$ defined on $X$ with $\int_X |f(x)|^p \mu(dx) < \infty$. Define a mapping $(f, t) \to f_t$ of $L_p \times R$ into $L_p$ by

$$f_t(x) = f(\pi(x, t)) \ .$$

This mapping does define a global flow on $L_p$.

Let $U_t f = f_t$. We claim that $U_t$ is a bounded linear operator on $L_p$. In fact

$$\|U_t f\|^p = \int_X |f(\pi(x, t))|^p \mu(dx) \ .$$

By the change of variables $y = \pi(x, t)$, the last equality becomes

$$\|U_t f\|^p = \int_X |f(y)|^p \mu(dy) = \|f\|^p ,$$

where we use the fact that $\mu$ is invariant with respect to $\pi$. Hence $U_t$ is an isometry for all $t$. Also we note that $U_t U_s = U_{t+s}$.

For $\tau > 0$, define a mapping* $A_\tau : L_p \to L_p$ by

$$A_\tau f(x) = \frac{1}{\tau} \int_0^\tau f(\pi(x, s))ds .$$

$A_\tau$ is the time-average of $f$ over the interval $[0, \tau]$. First we shall show that the range of $A_\tau$ is indeed in $L_p$.

$$\|A_\tau f\|^p = \int_X |A_\tau f(x)|^p \mu(dx) = \int_X \left| \frac{1}{\tau} \int_0^\tau f(\pi(x, s))ds \right|^p \mu(dx) .$$

By the Minkowsky inequality we get

$$\|A_\tau f\|^p \le \int_X \frac{1}{\tau^p} \int_0^\tau |f(\pi(x, s))|^p ds \cdot \left( \int_0^\tau 1^q ds \right)^{p/q} \mu(dx)$$

where $1/p + 1/q = 1$. Hence

$$\|A_\tau f\|^p \le \left(\frac{1}{\tau}\right)^p (\tau)^{p/q} \int_X \int_0^\tau |f(\pi(x, s)|^p ds \mu(dx) .$$

By interchanging the order of integration we get

$$\|A_\tau f\|^p \le \tau^{-1} \int_0^\tau \int_X |U_s f(x)|^p \mu(dx) = \tau^{-1} \int_0^\tau \|f\|^p ds = \|f\|^p.$$

---

* We use here the notation that is appropriate when the group $T$ is the reals $R$. If $T = I$, the integral from 0 to $\tau$ becomes a summation.

Hence $\|A_\tau\| \leq 1$.

We can now state the ergodic theorem.

X.1 THEOREM (Birkhoff Ergodic Theorem I.) *Let* $\pi$ *be a global flow on a uniform space* $X$ *and assume that there exists a normalized invariant measure* $\mu$. *Then for* $1 \leq p < \infty$ *there is a bounded linear mapping* $Q: L_p(X,\mu) \to L_p(X,\mu)$ *with the following properties: If* $F = Qf$, *then*

(i) $$\lim_{\tau \to \infty} \frac{1}{\tau} \int_0^\tau f(\pi(x,t))dt = F(x) ,$$

*where the limit is taken in the norm on* $L_p$,

(ii) $U_t F = F$, *for all* $t$,

(iii) $$\int_X f(x)\mu(dx) = \int_X F(x)\mu(dx) .$$

Before we prove this, let us note that the limit in (i) can be replaced by the following:

(i′) $$\lim_{\sigma,\tau \to \infty} \frac{1}{\tau+\sigma} \int_{-\sigma}^{\tau} f(\pi(x,t))\,dt = F(x) .$$

*Proof:* We shall prove this for the space $L_2(X,\mu)$, which is a real Hilbert space. A proof for the general case can be found in [9, pp. 684 ff].

The first thing we want to do is examine the limit in (i), which is $\lim_{\tau \to \infty} A_\tau f$. As a first step in this, let us restrict $\tau$ to the integers. Then

$$A_n f(x) = \frac{1}{n} \int_0^n f(\pi(x, t))dt = \frac{1}{n} \sum_{m=0}^{n-1} \int_m^{m+1} f(\pi(x, t))dt$$

$$= \frac{1}{n} \sum_{m=0}^{n-1} \int_0^1 f(\pi(x, r+m))dt = \frac{1}{n} \sum_{m=0}^{n-1} \int_0^1 f(\pi(\pi(x, m), t)dt$$

$$= \frac{1}{n} \sum_{m=0}^{n-1} U_m\left( \int_0^1 f(\pi(\cdot\,, t))dt \right)(x) \,.$$

That is $A_n f = 1/n\, \Sigma_{m=0}^{n-1} U_m A_1 f = B_n A_1 f$, where $B_n = 1/n\, \Sigma_{m=0}^{n-1} U_m$.

Let $M = \overline{\mathcal{R}(A_1)}$ be the closure of the range of $A_1$. $M$ is then a Hilbert space. Let

$$M_1 = \mathfrak{N}(I - U_1) \cap M = \{g \in M\colon\; U_1 g = g\}$$

then $M = M_1 \oplus M_2$ where $M_2 = M_1^{\perp}$ is the orthogonal complement of $M_1$ in $M$. Since $M_1$ is an invariant subspace of $U_1$ it follows that $M_2$ is an invariant subspace of the adjoint $U_1^*$. Since $U_1$ is a unitary mapping one has $U_1^* = U_1^{-1}$ (cf. [35]). Hence $M_2$ is an invariant subspace of $U_1$.

If $g_1 \in M_1$. then $U_1 g_1 = g_1$ and $B_n g_1 = g_1$. Hence $B_n g_1 \to g_1$ as $n \to \infty$.

Since the range of $(I - U_1)$ is dense in $M_2$, it follows that any function $g_2$ in $M_2$ can be written in the form

$$g_2 = \hat{g}_2 - U_1 \hat{g}_2 + h_2$$

where $\|h_2\| < \varepsilon$ and $\varepsilon$ is any preassigned positive number. Then

$$B_n g_2 = B_n(\hat{g}_2 - U_1 \hat{g}_2) + B_n h_2$$

$$= \frac{1}{n}[\hat{g}_2 - U_n \hat{g}_2] + B_n h_2 \,.$$

Hence

$$\|B_n g_2\| \leq \frac{2\|\hat{g}_2\|}{n} + \|B_n\| \cdot \|h_2\| \leq \frac{2\|\hat{g}_2\|}{n} + \varepsilon .$$

If we let $n \to \infty$ we get

$$\limsup_{n\to\infty} \|B_n g_2\| \leq \varepsilon .$$

Since $\varepsilon$ is arbitrary, we have $B_n g_2 \to 0$ as $n \to \infty$.

If we let $P$ denote the orthogonal projection of $L_2(X, \mu)$ onto $M_1$ is is clear that $B_n g \to Pg$ as $n \to \infty$, for every $g$ in $M$. Hence $A_n f \to PA_1 f$ for every $f$ in $L_2(X, \mu)$. The mapping $Q$ is then $Q = PA_1$. Since $\|Q\| \leq \|P\| \cdot \|A_1\| \leq 1$, we see that $Q$ is bounded.

If we consider $A_\tau$ when $\tau$ is not an integer, then one has $\tau = n + r$ for some $r$, $0 \leq r < 1$. One can then show that

$$(A_\tau - A_n)f = \frac{r}{n+r}[A_n + U_n A_r]f .$$

Hence

$$\|(A_\tau - A_n)f\| \leq \frac{r}{n+r} (\|A_n\| + \|U_n\| \|A_r\|)\|f\| \leq \frac{r}{n+r} 2\|f\| \to 0$$

as $\tau \to \infty$. Therefore $\lim_{\tau\to\infty} A_\tau f = \lim_{n\to\infty} A_n f = PA_1 f$. (One can similarly show that the limit in $(i')$ is as indicated.)

Since

$$U_t F(x) = F(\pi(x, t)) = \lim_{\tau,\sigma\to\infty} \frac{1}{\tau+\sigma} \int_{-\sigma}^{\tau} f(\pi(\pi(x, t), s)\, ds$$

$$= \lim_{\tau,\sigma\to\infty} \frac{1}{\tau+\sigma} \int_{-\sigma}^{\tau} f(\pi(x, t+s))\, ds$$

$$= \lim_{\tau,\sigma\to\infty} \frac{1}{\tau+\sigma} \int_{t-\sigma}^{t+\tau} f(\pi(x, s))\, ds = F(x),$$

we see that (ii) holds.

Finally, since integration over $X$ is a continuous linear functional one has

$$\int_X F(x)\mu(dx) = \int_X Qf\,\mu(dx) = Q\int_X f(x)\mu(dx)$$

$$= \lim_{T\to\infty} A_T \int_X f(x)\,\mu(dx) = \lim_{T\to\infty} \frac{1}{T}\int_0^T \ell(f)\,dt = \ell(f)$$

where $\ell(f)$ is the constant function $\ell(f) = \int_X f(x)\mu(dx)$. This proves (iii). QED

One can get further information concerning the limit function $F(x)$ in (i) or (i′). For this we need another concept.

A flow $\pi$ is said to be **irreducible** (with respect to an invariant measure $\mu$) or $\mu$-**irreducible** if for every measurable invariant set $A \subset X$ one has

$$\mu(A)\cdot\mu(X-A) = 0\,,$$

that is either $A$ or its complement has $\mu$-measure zero.

X.2 THEOREM. (Birkhoff Ergodic Theorem II.) *Let $\pi$ be a global flow on a uniform space $X$ and assume that there is a normalized invariant measure $\mu$ where $\pi$ is $\mu$-irreducible. Then the function $F$ in Theorem X.1 is the constant function*

$$F(x) = \int_X f\mu(dx)\ .$$

*In particular, one has*

$$\int_X f(x)\mu(dx) = \lim_{T\to\infty} \frac{1}{T}\int_0^T f(\pi(x,t))\,dt\ ,$$

*that is, the time-average of* $f(\pi(x,t))$ *and the space-average of of* $f(x)$ *agree.*

*Proof:* Let $f \in L_p(X,\mu)$ and let $\alpha = \int_X f(x)\mu(dx)$. Since $\mu(X) = 1$, it follows that $\alpha$ is finite. Let

$$A = \{x \in X : F(x) < \alpha\}$$

$$B = \{x \in X : \alpha < F(x)\}$$

where $F = Qf$ is given by Theorem X.1. Since $U_t F = F$ for all $t$, it follows that both $A$ and $B$ are invariant subsets of the flow $\pi$ on $X$. It follows from the $\mu$-irreducibility that either $A$ or $X - A$ has measure 0, and similarly $B$ or $X - B$ has measure zero. Since $\int_X F(x)\mu(dx) = \alpha$, it follows that both $A$ and $B$ have measure 0. Hence $F(x) = \alpha$, $\mu$-almost everywhere. QED

X.B **Ergodic Theory on** $W \times H(f)$. Throughout this section we shall assume that $f$ is a regular function in $C(W \times R, R^n)$ and that the motion $f_t$ is compact. We shall let $\mu$ be a normalized invariant measure for the flow on $H(f)$. If $f_t$ is almost periodic, $\mu$ would be the Haar measure. Furthermore, we shall assume that the flow $\pi(x, f, t)$ on $W \times H(f)$ is a global flow.

X.3. THEOREM. *Assume there is a normalized measure* $\nu$ *on* $W$ *that is invariant with respect to* $f$. *Let* $m = \nu \times \mu$ *be the product measure on* $W \times H(f)$. *Then for every* $g \in L_p(W,\nu)$, *where* $1 \le p < \infty$, *there is a* $G \in L_p(W \times H(f), m)$ *such that*

(i) $$\lim_{T \to \infty} \frac{1}{T} \int_0^T g(\phi(x,f,t))\,dt = G(x,f)$$

(ii) $$G(\pi(x,f,t)) = G(\phi(x,f,t), f_t) = G(x,f)$$

(iii) $$\int_W g\, d\mu = \int_{W \times H(f)} G\, dm$$

*Proof:* It follows from Theorem IX.7 that $m$ is a normalized invariant measure for $\pi$. This theorem now follows from the Birkhoff Ergodic Theorem I while noting that if $g \in L_p(W, \nu)$ and $\hat{g}$ is defined by

$$\hat{g}(x, f) = g(x) \qquad ((x, f) \in W \times H(f))$$

then $\hat{g} \in L_p(W \times H(f), m)$, $\hat{g}(\pi(x, f, t) = g(\phi(x, f, t))$ and

$$\int_{W \times H(f)} \hat{g}\, dm = \int_W g\, d\nu$$

QED

An interesting problem is to decide whether the function $G$ in Theorem X.3 is actually independent of $f$, which would mean that we could view $G$ an an element of $L_p(W, \nu)$. This seems like a hard problem. All we are able to do here is to show that, under certain conditions on $\mu$, there exist functions $G$ that are independent of $f$.

X.4 THEOREM. *Assume, in addition to the hypotheses of Theorem X.3, that the flow $f_t$ is $\mu$-irreducible on $H(f)$. Let $g$ and $G$ be related by Theorem X.3. Then there is a $h$ in $L_p(W, \nu)$ such that*

(i) $$\lim_{T \to \infty} \frac{1}{T} \int_0^T G(x, f_s)\, ds = h(x)$$

(ii) $$h(\phi(x, f, t)) = h(x)$$

(iii) $$\int_W g\,d\nu = \int_W h\,d\nu \;.$$

*Proof:* Since the flow $f_t$ is $\mu$-irreducible on $H(f)$, it follows from Birkhoff's Second Ergodic Theorem, that the limit in (i) is independent of $f$. Equalities (ii) and (iii) follow from the counterparts in Theorem X.3. QED

The relationship $\Phi$ that maps $g$ onto $h$ by Theorems X.3 and X.4 is a bounded linear operator from $L_p(W, \nu)$ into itself. It is given by

$$h(x) = \lim_{\sigma,\tau \to \infty} \frac{1}{\sigma\tau} \int_0^{\sigma} \int_0^{\tau} g(\phi(x, f_s, t))\,dt\,ds \;.$$

*Exerice 1.* One application of Theorem X.4 occurs when the flow $f_t$ is almost periodic. Prove this. That is, let $E$ be an a.p. minimal set in a flow $\pi$ and let $\mu$ be the Haar measure on $E$. Show that $\pi$ is $\mu$-irreducible.

# CHAPTER XI

## *APPLICATIONS TO INTEGRAL EQUATIONS*

In this chapter we shall show how the solutions of certain integral equations give rise to semiflows. We shall see that the semi flow we construct is actually an extension of the flow $\pi(x, f, t)$ constructed in Chapter IV.

We will not attempt to give a complete discussion of the theory of this semiflow. More details can be found in [19] and [20]. In this chapter we shall concentrate on the basic construction of the semiflow and give one application.

XI.A. **Volterra integral equations.** Consider now the Volterra integral equation

$$x(t) = f(t) + \int_0^t a(t, s) g(x(s), s) ds \quad , \tag{1}$$

where $x$ is an $n$-vector, $f: R^+ \to R^n$, $g: R^n \times R^+ \to R^n$ and $a$ is an $n \times n$ matrix. If $\phi(t) = \phi(f, g, a; t)$, $0 \le t < \alpha$, is a solution of (1) we shall define the function $T_\tau f = T_\tau(f, g, a)$ by

$$(T_\tau f)(\theta) = f(\tau + \theta) + \int_0^\tau a(\tau + \theta, s) g(\phi(s), s) ds \tag{2}$$

for $0 \le \tau < \alpha$ and $0 \le \theta$. Define $g_\tau$ and $a_\tau$ by

$$g_\tau(x, s) = g(x, \tau + s), \qquad a_\tau(t, s) = a(\tau + t, \tau + s) \quad .$$

Now define a mapping $\pi$ (formally) by

(3) $$\pi(f, g, a; \tau) = (T_\tau f, g_\tau, a_\tau) \;,$$

where $0 \leq \tau < \alpha$. We will show shortly that the mapping $\pi$ defines a semiflow on a space $X$ that consists of ordered triples $(f, g, a)$.

Let us be more precise about (1). First we shall assume that the function $f$ lies in $C = C(R^+, R^n)$, cf. Example III.E. Next we shall assume that the function $g$ lies in $\mathfrak{L}_p = \mathfrak{L}_p(R^n \times R^+, R^n)$, cf. Example III.N and that the kernel $a(t, s)$ lies in $\mathfrak{A}^*_p$, cf. Example III.O, where $1 \leq p < \infty$. We assume that $a(t, s)$ is defined only for $0 \leq s \leq t < \infty$, or equivalently, that $a(t, s) = 0$ for $t < s$.

The next result, an existence theorem, can be proved by using the contraction mapping theorem. (Recall that the functions $g(x, t)$ in $\mathfrak{L}_p$ are Lipschitz-continuous in $x$.) Also see [19].

XI.1. THEOREM. *Let $f \in C$, $g \in \mathfrak{L}_p$ and $a \in \mathfrak{A}^*_p$, where $1 \leq p < \infty$. Then there is precisely one solution $\phi(t) = \phi(f, g, a; t)$ of* (1) *defined on the maximal interval $0 \leq t < \alpha$. Furthermore, either $\alpha = +\infty$ or $\phi(t) \to \infty$ as $t \to \alpha$.*

As a consequence of this existence *and* uniqueness theorem the mapping $T_\tau f$ in (2) is defined for all $\tau$, $0 \leq \tau < \alpha$. and for all $\theta \geq 0$. Furthermore we can now show that the mapping $\pi$ in (3) has the semigroup property, or $\pi$ satisfies axiom (ii) in the definition of a flow.

We want to show that

(4) $$\pi(T_\tau f, g_\tau, a_\tau; \sigma) = (T_{\tau+\sigma} f, g_{\tau+\sigma}, a_{\tau+\sigma})$$

where $\tau \geq 0$ and $\sigma \geq 0$. The left side of (4) suggests that we want to solve the integral equation

(5) $$x(t) = (T_\tau f)(t) + \int_0^t a_\tau(t, s) g_\tau(x(s), s)\, ds$$

and then compute $T_\sigma(T_\tau f)$ by (2). Let $\psi(t)$, $0 \leq t \leq \sigma$, be the solution of (5), and $\phi(s)$, $0 \leq s < \sigma$, the solution of (1). Since

$$\begin{aligned}
\phi(t+\tau) &= f(t+\tau) + \int_0^{t+\tau} a(t+\tau, s) g(\phi(s), s)\, ds \\
&= f(t+\tau) + \int_0^{\tau} + \int_\tau^{t+\tau} [a(t+\tau, s) g(\phi(s), s)\, ds] \\
&= (T_\tau f)(t) + \int_0^t a(\tau + t, \tau + s) g(\phi(\tau + s), \tau + s)\, ds \\
&= (T_\tau f)(t) + \int_0^t a_\tau(t, s) g_t(\phi(\tau + s), s)\, ds ,
\end{aligned}$$

the uniqueness of solutions implies that $\phi(t+\tau) = \psi(t)$ for $0 \leq t \leq \sigma$. Next we have

$$\begin{aligned}
T_\sigma(T_\tau f)(\theta) &= (T_\tau f)(\sigma + \theta) + \int_0^\sigma a_\tau(\sigma + \theta, s) g_\tau(\psi(s), s)\, ds \\
&= f(\tau + \sigma + \theta) + \int_0^\tau a(\tau + \sigma + \theta, s) g(\phi(s), s)\, ds \\
&\quad + \int_\tau^{\tau+\sigma} a_\tau(\sigma + \theta, s-\tau) g_\tau(\psi(s-\tau), s-\tau)\, ds \\
&= f(\tau + \sigma + \theta) + \int_0^{\tau + \sigma} a(\tau + \sigma + \theta, s) g(\phi(s), s)\, ds \\
&= (T_{\tau+\sigma} f)(\theta) .
\end{aligned}$$

Since the mapping $(g, \tau) \to g_\tau$ and $(a, \tau) \to a_\tau$ satisfy the semi-group property, we see that $\pi$ does as well.

The next question is continuity and for this we impose topologies on C, $\mathfrak{L}_p$ and $\mathfrak{A}_p^*$. On C we shall use the topology of uniform convergence on compact sets. We shall use the topology $\mathfrak{T}_b$ on $\mathfrak{L}_p$, cf. Example III.N. And on $\mathfrak{A}_p^*$ we use the topology $\mathfrak{T}_\omega$, cf. Example III.O. The following continuity result is proved in [19]. (Also see Appendix A.)

XI.2. THEOREM. *Assume that the spaces C, $\mathfrak{L}_p$ and $\mathfrak{A}_p^*$ have the prescribed topologies. Then the solution $\phi(f, g, a; t)$ of* (1) *depends continuously on the four variables. More precisely, if $f_n \to f$ in $C(R^+, R^n)$, $g_n \to g$ in $(\mathfrak{L}_p, \mathfrak{T}_b)$, $a_n \to a$ in $(\mathfrak{A}_p^*, \mathfrak{T}_\omega)$ and $t_n \to t$ in $R^+$. where $t_n$ is a point in the maximal interval of definition of the solution $\phi(f_n, g_n, a_n; t)$, then $t$ lies in the interval of definition of $\phi(f, g, a; t)$ and*

$$\phi(f_n, g_n, a_n; t_n) \to \phi(f, g, a; t) \ .$$

One consequence of this theorem is that if $[0, \alpha_n)$ denotes the maximal interval of definition of $\phi(f_n, g_n, a_n; t)$, then $\alpha \leq \liminf \alpha_n$.

Another consequence of this theorem is that the mapping $T_\tau$ in (2), which actually depends on the four terms $f, g, a$ and $\tau$, is continuous in these variables. That is if $f_n \to f$, $g_n \to g$, $a_n \to a$, $\tau_n \to \tau$, then $T_{\tau_n} f_n(\theta)$ converges to $T_\tau f(\theta)$ uniformly for $\theta$ in compact sets in $R^+$.

A third consequence is the following:

XI.3. THEOREM. *Assume that the spaces C, $\mathfrak{L}_p$ and $\mathfrak{A}_p^*$ have the prescribed topologies. Then the mapping $\pi$ given by* (3) *defines a semiflow on $X = C \times \mathfrak{L}_p \times \mathfrak{A}_p^*$.*

XI.B **Abstract formulation for** $\pi(f, g, a; t)$. By recalling Section IV.F, the reader may suspect that the mapping $\pi$ may still define a flow on $X$ when $X$ has a different topology. As a matter of fact, many of the topological-dynamical properties of Volterra integral equations are not dependent on the specific topologies. In order to see this we need some definitions.

Assume that the terms $g(x, t)$ belong to a uniform topological space $\mathcal{G}$ and that the kernels $a(t, s)$ belong to a uniform topological space $\mathcal{A}$. We shall say that $\mathcal{G}$ is **admissible** if the mapping

$$(g, \tau) \to g_\tau$$

of $\mathcal{G} \times R^+$ into $\mathcal{G}$ is continuous. The space $\mathcal{A}$ is said to be **admissible** if the two mappings

$$(a, \tau) \to a_\tau \quad \text{and} \quad \tau \to a(\tau + \cdot\ , \cdot),$$

of $\mathcal{A} \times R^+$ into $\mathcal{A}$ and $R^+$ into $\mathcal{A}$, are continuous. The pair $(\mathcal{G}, \mathcal{A})$ is said to be **compatible** if

(i) both $\mathcal{G}$ and $\mathcal{A}$ are admissible and

(ii) for each $f \in C$, $g \in \mathcal{G}$, and $a \in \mathcal{A}$, the Volterra integral equation

$$x(t) = f(t) + \int_0^t a(t, s) g(x(s), s)\, ds$$

admits a unique solution $\phi(f, g, a; t)$ on some maximal interval $0 \leq t < \alpha$, and moreover, this solution depends continuously on $f, g, a$ and $t$ in the precise sense as given in Theorem XI.2.

XI.4. THEOREM. *Let* $(\mathcal{G}, \mathcal{A})$ *be a compatible pair of spaces. Let* $\phi(t) = \phi(f, g, a; t)$ *be the unique solution of*

$$x(t) = f(t) + \int_0^t a(t, s)\, g(x(s), s)\, ds \quad ,$$

*where* $f \in C$, $g \in \mathcal{G}$ *and* $a \in \mathcal{A}$, *on the maximal interval* $[0, \alpha)$. *Then the mapping* $\pi$ *given by* (3) *is a semiflow on* $C \times \mathcal{G} \times \mathcal{A}$, *where the interval of definition of the motion* $\pi(f, g, a; t)$ *is* $[0, \alpha)$.

In the last section we asserted that the spaces $\mathcal{G} = (\mathcal{L}_p, \mathcal{T}_b)$ and $\mathcal{A} = (\mathcal{A}_p^*, \mathcal{T}_\omega)$ are compatible. Other examples of compatible pairs are given in [20]. We note here that the spaces $\mathcal{G} = (\mathcal{L}_p, \mathcal{T}_c)$ and $\mathcal{A} = (\mathcal{A}_p^*, \mathcal{T}_s)$ are also compatible. It is interesting to observe that while one weakens the topology on $\mathcal{L}_p$ (from $\mathcal{T}_b$ to $\mathcal{T}_c$) one strengthens the topology on $\mathcal{A}_p^*$ (from $\mathcal{T}_\omega$ to $\mathcal{T}_s$) in order to preserve compatibility.

XI.C. **The mapping** $T_\tau f$. If one sets $\theta = 0$, then Equation (2) becomes

$$T_\tau f(0) = \phi(\tau) = \phi(f, g, a; \tau) \quad ,$$

in other words, $T_\tau f(0)$ agrees with the solution of (1). Secondly, if the kernel $a(t, s)$ depends only on $s$ and if $f$ is constant, then $T_\tau f(\theta)$ is a constant function, that is

$$T_\tau f(\theta) = \phi(\tau)$$

for all $\theta \geq 0$. This situation occurs precisely when one changes the initial value problem of the differential equation

$$x' = g(x, t), \qquad x(0) = x_0$$

into the integral equation

$$x(t) = x_0 + \int_0^t g(x(s), s)\,ds \ .$$

Consequently the semiflow $\pi(f, g, a; \tau)$ in (3) is an extension of the (semi) flow defined in Section IV.F.

The following reformulations of $T_\tau f$ can be established by means of a routime change of variables.

(6) $$T_\tau f(\theta) = f_\tau(\theta) + \int_{-\tau}^{0} a_\tau(\theta, s)\, g_\tau(\phi_\tau(s), s)\,ds$$

(7) $$T_\tau f(\theta) = \phi(\tau + \theta) - \int_0^{\theta} a_\tau(\theta, s)\, g_\tau(\phi_\tau(s), s)\,ds \ .$$

Equation (6) is always valid, whereas Equation (7) is valid only when the solution $\phi(t)$ can be continued to $t = \tau + \theta$.

XI.D. **Limiting Equations.** Let $(\mathcal{G}, \mathfrak{A})$ be a compatible pair of spaces and let $\pi$ be given by (3). Let $\Omega$ denote the $\omega$-limit set of the motion $\pi(f, g, a; t)$ in $X = C \times \mathcal{G} \times \mathfrak{A}$. The collection of all Volterra integral equations of the form

(8) $$X(t) = F(t) + \int_0^t A(t, s)\, G(X(s), s)\,ds$$

where $(F, G, A) \in \Omega$ is said to be the **set of limiting equations of**

(9) $$x(t) = f(t) + \int_0^t a(t, s)\, g(x(s), s)\,ds \ .$$

One can give a comparison between the solutions (9) and those of (8) much in the same spirit as we did in Section IV.D. For example, if the motion $\pi(f, g, a; t)$ in $X$ is positively compact, then the of limiting equations is a nonempty compact connected subset of $X$. Another comparison is the following:

XI.5. THEOREM. *Let* $(F, G, A) \in \Omega$, *where* $\Omega$ *is the* $\omega$*-limit set of the motion* $\pi(f, g, a; t)$. *Then there is a sequence* $\{\tau_n\}$ *with* $\tau_n \to \infty$ *and*

$$\phi(f, g, a; \tau_n + t) \to \phi(F, G, A; t)$$

*where the convergence is uniform on compact subsets of* $[0, \alpha)$, *the maximal interval of definition of* $\phi(F, G, A; t)$.

*Proof:* It follows from Proposition II.4 and II.6 that there is a sequence $\{\tau_n\}$ with $\tau_n \to \infty$ and

$$\pi(f, g, a; \tau_n + t) \to \pi(F, G, A; t)$$

uniformly on compact subsets of $[0, \alpha)$. In particular, this implies that

$$T_{\tau_n + t} f \to T_t F \tag{10}$$

uniformly for $t$ in compact subsets of $[0, \alpha)$. However, the convergence in (10) means that

$$T_{\tau_n + t} f(\theta) \to T_t F(\theta)$$

uniformly for $\theta$ in compact sets in $R^+$. If we set $\theta = 0$, this becomes

$$\phi(f, g, a; \tau_n + t) \to \phi(F, G, A; t) .$$

QED.

One of the problems that arises with this flow, that does not appear in the prototype in Chapter IV, is to describe the limiting behavior of $T_\tau f$, that is, to describe the function $F(t)$ in (8).

In an attempt to solve this we recall that the motion $a_\tau$ in $\mathfrak{A}$ is a fixed point if and only if the kernel $a(t, s)$ is of convolution-type (Proposition III.20), that is $a(t, s) = \alpha(t-s)$.

XI.6 THEOREM. *Let* $(\mathcal{G}, \mathfrak{A})$ *be a compatible pair and let* $g_t$ *be a positively compact motion in* $\mathcal{G}$ *and let* $a(t, s) = \alpha(t-s)$ *be in* $\mathfrak{A}$ *where* $\alpha$ *is in* $L_1[0, \infty)$. *Assume that for some* $f \in C$ *the motion* $f_t$ *is positively compact and the solution* $\phi(f, g, a; t) = \phi(t)$ *of*

$$(10) \qquad x(t) = f(t) + \int_0^t \alpha(t-s)\, g(x(s), s)\, ds$$

*is bounded and uniformly continuous. Then the limiting equations of* (10) *are of the form*

$$X(t) = f^*(t) + \int_{-\infty}^t \alpha(t-s)\, g^*(X(s), s)\, ds ,$$

*where* $X(t)$, $f^*(t)$ *and* $g^*(x, t)$ *are limits of* $\phi(t_n + t)$, $f(t_n + t)$ *and* $g(x, t_n + t)$ *for some sequence* $\{t_n\}$ *with* $t_n \to \infty$.

This means that the function $F$ that appears in (8) is of the form

$$(11) \qquad F(t) = f^*(t) + \int_{-\infty}^0 \alpha(t-s)\, g^*(X(s), s)\, ds .$$

*Proof:* Choose a sequence $\{t_n\}$ so that $t_n \to \infty$ and $g_{t_n} \to g^*$ (in $\mathcal{G}$), $f_{t_n} \to f^*$ (in $C$) and $\phi_{t_n}(t) \to X(t)$, where the last limit is uniform for $t$ in compact sets. (We use Theorem III.4 to get the last limit.) Then $T_{t_n} f$ as given by (6) becomes

$$T_{t_n} f(\theta) = f_{t_n}(\theta) + \int_{-t_n}^{0} a(\theta - s) g_{t_n}(\phi_{t_n}(s), s)\, ds$$

or,

$$T_{t_n} f(\theta) = f_{t_n}(\theta) + \int_{-\infty}^{0} a(\theta - s) g_{t_n}(\phi_{t_n}(s), s)\, ds$$

$$- \int_{-\infty}^{-t_n} a(\theta - s) g_{t_n}(\phi_{t_n}(s), s)\, ds \ .$$

If we let $t_n \to \infty$, the last equation becomes

$$F(\theta) = f^*(\theta) + \int_{-\infty}^{0} a(\theta - s) g^*(X(s), s)\, ds$$

$$- \lim_{t_n \to \infty} \int_{-\infty}^{-t_n} a(\theta - s) g_{t_n}(\phi_{t_n}(s), s)\, ds \ .$$

However, $g_{t_n}(\phi_{t_n}(s), s)$ is uniformly bounded. Since $a$ is in $L_1(0, \infty)$ the last limit above is zero. Hence $F$ is given by (11). QED.

XI.E. **An application.** Consider the real scalar equation

$$x(t) = f(t) + \int_0^t a(t - s) g(x(s))\, ds \ , \tag{12}$$

where $f(t) \to f_0$ as $t \to \infty$, the function $a(r)$ is a $C^1$ function for $0 \leq r < \infty$ and $a(r) \in L_p(0, \infty)$. Assume further that $a(r) \leq 0$, $a'(r) \geq 0$ and $a'(r) \neq 0$ in any interval $[0, \tau]$, $\tau > 0$. Assume that $g(x)$ is locally Lipschitzian in $x$, and strictly increasing in

Consider the equation

$$x = f_0 + Ag(x) \quad ,$$

where $A = \int_0^\infty a(r)\,dr$. This equation has a unique solution, which we denote by $x_0$. Now assume that the solution $\phi(t)$ of (12) is defined for all $t \geq 0$ and that this solution is bounded. (See Exercise 1.) We claim that $\phi(t) \to x_0$ as $t \to \infty$.

If we let $y(t) = \phi(t) - f(t) - x_0 + f_0$, then $y(t)$ satisfies the integral equation

$$(13) \quad y(t) = \left(-\int_t^\infty a(s)\,ds\right)g(x_0) + \int_0^t a(t-s)h(y(s), s)\,ds$$

where $h(y, s) = g(y + f(s) - f_0 + x_0) - g(x_0)$. It follows from Theorem XI.6 that the limit equation of (13) (there is only one) has the form

$$(14) \qquad Y(t) = \int_{-\infty}^t a(t-s)[g(Y(s) + x_0) - g(x_0)]\,ds \ .$$

We shall now outline the remainder of the argument. We ask the reader to check the details.

Let $y_0 = \liminf y(t)$ and $y_1 = \limsup y(t)$, as $t \to \infty$. One can easily show that $y_0 \leq 0 \leq y_1$. In order to show that $y_1 = 0$ we pick a sequence $\{\tau_n\}$ so that $\tau_n \to \infty$ and

$$y(\tau_n) \to y_1 \text{ and } y(\tau_n + t) \to Y(t)$$

where $Y(t)$ is given by (14). It follows that $Y(0) = y_1 \geq Y(t)$ for all $t$ in $R$.

Now assume that $y_1 > 0$. If $Y(t) = y_1$ for all $t \leq 0$, then

$$0 < y_1 = Y(0) = \int_{-\infty}^{0} a(-s)[g(y_1 + x_0) - g(x_0)] \leq 0,$$

which is a contradiction. If $Y(-T) = y_1$ and $Y(t) \neq y_1$ for $t$ in the interval $-S \leq t < -T$ (where $S > 0$) then one can show that

$$Y'(-T) < \left[a(0) - \int_0^{\infty} a'(s)\,ds\right] \cdot [g(y_1 + x_0) - g(x_0)] = 0 .$$

This shows that $Y(t) > Y(-T) = y_1$ in a small interval to the left of $-T$, which is a contradiction. Hence $y_1 = 0$. Similarly one can show that $y_0 = 0$. It follows that $\phi(t) \to x_0$ as $t \to \infty$.

We have presented only one application here. The reader who is interested in more information is invited to read [20].

*Exercise* 1. Let $g(x)$ be a continuous function with Lipschitz coefficient $k$, that is, $|g(x) - g(y)| \leq k|x - y|$. Assume that $g(0) = 0$. Let $a(r)$ be an $L_1(0,\infty)$ function with $A = \int_0^{\infty} |a(r)|\,dr$. Assume that $Ak < 1$. Let $f(t)$ be a bounded continuous function and let $\|f\| = \sup\{|f(t)|: t \geq 0\}$. Show that the integral equation

$$x(t) = f(t) + \int_0^{t} a(t-s)g(x(s))\,ds$$

has a bounded solution $x(t)$ with $\|x\| \leq (1 - Ak)^{-1}\|f\|$.

# APPENDIX A.

## *KAMKE'S LEMMA*

In several examples of flows and semiflows it was necessary to verify a continuity property similar to that appearing in Kamke's Lemma (Lemma IV.2). This question of continuity appeared in the discussion of admissibility in Section IV.F and the discussion of compatibility in Section XI.B.

In this appendix we shall outline two proofs of Kamke's lemma. We do this in order to familiarize the reader with the two basic (and different) techniques for settling the question of continuity. The first proof we give is based on a compactness argument and the second proof is based on a differential inequality.

A.1 **Compactness Proof.** We shall use the notation of Lemma IV.2. The object of this argument is to show that the sequence of solutions $\{\phi(x_n, f_n; t\}$ is a bounded, equicontinuous family of continuous functions when $|t| \leq \sigma$. One can then use Ascoli's Theorem to construct a convergent subsequence. (It is then a simple matter to verify the limit function $\psi(t)$ is a solution of the limiting equation $x' = f(x, t)$ and by uniqueness one has $\psi(t) = \phi(x, f, t)$.) This, of course, leads to convergence only for $t$ satisfying $|t| \leq \sigma$. In order to get convergence at $t = \tau$, we simply repeat the first part thereby enlarging the $t$-interval to $|t| \leq \sigma'$ $(\sigma' > \sigma)$. Because of compactness one will reach the time $t = \tau$ after a finite number of such steps.

Let us now outline the proof that $\{\phi(x_n, f_n, t)\}$ is bounded and equicontinuous for $|t| \leq \sigma$, at least for $n$ sufficiently large. Let $K$ be a compact neighborhood of the solution $\phi(x_0, f, t)$, where $x_n \to x_0$, $f_n \to f$. Since $f_n \to f$ in $C(W \times R, R^n)$ it follows that for $n$ sufficiently large, the functions $f_n$ and $f$ have a common bound on $K$, that is

$$|f_n(x, t)| \leq M, \qquad |f(x, t)| \leq M, \qquad ((x, t) \in K).$$

Now choose $\alpha, \beta$ $(0 < \alpha < \beta)$ so that the sphere

$$S(x_0, \beta) = \{x \in R^n : |x - x_0| < \beta\}$$

lies in $K$. Then $x_n \in S(x_0, \alpha)$ for $n$ sufficiently large and

$$\phi(x_n, f_n, t) \in \mathrm{Cl}\, S(x_0, \beta) \subset K$$

for $|t| \leq \sigma$ where $\sigma = (\beta - \alpha)/M$. It follows that for $n$ sufficiently large the functions $\phi(x_n, f_n, t)$ are bounded for $|t| \leq \sigma$. They are also equicontinuous since

$$|\phi(x_n, f_n, t) - \phi(x_n, f_n, s)| = \left| \int_s^t f_n(\phi_n(x_n, f_n, \xi), \xi) d\xi \right|$$

$$= M\,|t - s| \;.$$

A.2 **Differential Inequality Proof.** This method of argumentation can be used when the limit function $f$ satisfies a Lipschitz condition. This method is not as widely applicable as the compactness method, but it leads to the conclusion much faster.

Again we can assume that the solutions $\phi(x_n, f_n, t)$ and $\phi(x_0, f, t)$ lie in a fixed compact set $K$ for $|t| \leq \sigma$. Let $k$ be a

Lipschitz constant for $f$ on $K$, and let $\phi_n(t) = \phi(x_n, f_n, t)$ and $\phi(t) = \phi(x_0, f, t)$. Then for $t \geq 0$ one has

$$|\phi_n(t) - \phi(t)| \leq |x_n - x_0| + \int_0^t |f_n(\phi_n(s), s) - f(\phi(s), s)|\, ds$$

$$\leq |x_n - x_0| + \int_0^t |f_n(\phi_n(s), s) - f(\phi_n(s), s)|\, ds$$

$$+ \int_0^t |f(\phi_n(s), s) - f(\phi(s), s)|\, ds$$

$$\leq \varepsilon_n + \int_0^t k\, |\phi_n(s) - \phi(s)|\, ds$$

where $\varepsilon_n = |x_n - x_0| + \int_0^t |f_n(\phi_n(s), s) - f(\phi_n(s), s)|\, ds$. Since $x_n \to x_0$ and $f_n \to f$ (uniformly on $K$) one has $\varepsilon_n \to 0$. By Gronwall's inequality [8], one has

$$|\phi_n(t) - \phi(t)| \leq \varepsilon_n \exp kt \quad ,$$

hence $\phi_n(t) \to \phi(t)$ uniformly for $|t| \leq \sigma$.

If the reader would like to see these two methods of argumentation used for integral equations he is invited to read [19].

## *REFERENCES*

This is an abbreviated list of references on the subject matter. For an extensive and complete bibliography the reader is referred to [12] below.


[1] J. Auslander, H. Bhatia and P. Seibert, *Bol. Soc. Mat. Mexicana* (2) 9 (1964), 55-66.

[2] L. Auslander, F. Hahn and L. Markus, *Flows on Homogeneous Spaces*, pp. 43-58, Ann. Math. Studies No. 53, Princeton Univ. Press, Princeton, N.J., 1963.

[3] L. Auslander and F. Hahn, *Trans. Amer. Math. Soc.*, 106 (1963), 415-426.

[4] J. D. Baum, *Proc. Amer. Math. Soc.*, 4 (1953), 656-662.

[5] N. Bourbaki, *Elélements de Mathématique. Topologie Générale*, Hermann, Paris, 1965.

[6] D. Bushaw, *Differential and Functional Equations*, pp. 13-26, Benjamin, New York, 1967.

[7] J. Cronin, *Fixed Points and Topological Degree in Nonlinear Analysis*, Math. Surveys, No. 11, Amer. Math. Soc., Providence, R.I., 1964.

[8] E.A.Coddington and N. Levinson, *Theory of Ordinary Differential Equations*, McGraw-Hill, New York, 1955.

[9] N. Dunford and J. T. Schwartz, *Linear Operators, Part I*, Interscience, New York, 1963.

[10] J. Favard, *Leçons sur les fonctions presque périodiques*, Gauthier-Villars, Paris, 1933.

[11] W. Feller, *An Introduction to Probability Theory and its Applications*, Vol. 1, Wiley, New York, 1957.

[12] W. Gottschalk, *Bibliography for Topological Dynamics*, 3 ed., Wesleyan Univ., Middletown, Conn., 1967.

[13] W. Gottschalk and G. Hedlund, *Topological Dynamics*, Amer. Math. Soc. Colloq. Publ., Vol. 36, Providence, R. I. , 1955.

[14] F. Hausdorff, *Fund. Math.*, 6 (1924), 146-148.

[15] J. Kelley, *General Topology*, Van Nostrand, Princeton, N.J., 1955.

[16] J. LaSalle, *Proc. Sympos. Appl. Math.*, Vol. 13, pp. 299-307. Amer. Math. Soc., Providence, R. I., 1962.

[17] L. Loomis, *Introduction to Abstract Harmonic Analysis*, Van Nostrand, Princeton, N. J., 1953.

[18] L. Markus, *Contributions to Nonlinear Oscillations*, Vol. 3, Princeton, Univ. Press, Princeton, N. J., 1956.

[19] R. K. Miller and G. R. Sell, *Existence, uniqueness and continuity of solutions of integral equations* (to appear in Ann. di Mat. Pura ed Appl.9

[20] R. K. Miller and G. R. Sell, *Volterra integral equations and topological dynamics*, (to appear).

[21] V. V. Nemyckiǐ, *Topological problems of the theory of dynamical systems*, Amer. Math. Soc. Transl. (1) 5 (1962), pp. 414-497.

[22] V. V. Nemyckiǐ, and V. V. Stepanov, *Qualitative Theory of Differential Equations*, Princeton Univ. Press, Princeton, N. J. 1960.

[23] Z. Opial, *Ann. Polon. Math.*, 8 (1960), 75-89.

[24] E. Roxin, *J. Differential Equations,* 1 (1965), 188-205.

[25] H. L. Royden, *Real Analysis,* MacMillan, New York, 1963.

[26] G.R. Sell, *J. Differential Equations,* 1 (1965), 370-392.

[27] G.R. Sell, *J. Differential Equations,* 2 (1966), 143-157-

[28] G.R. Sell, *Trans. Amer. Math. Soc.,* 127 (1967), 241-262.

[29] G.R. Sell, *Trans. Amer. Math. Soc.* 127 (1967), 263-283.

[30] G.R. Sell, *Compact sets of nonlinear operators* (to appear).

[31] G.R. Sell and Y. Sibuya, *Differential and Functional Equations,* pp. 501-506, Benjamin, New York, 1967.

[32] J. A. Yorke, *Mathematical Systems Theory,* 1 (1967), 353-372.

[33] T. Yoshizawa, *Stability Theory by Liapunov's Second Method,* Math. Soc. Japan, Tokyo, 1966.

[34] R. K. Miller, *Trans. Amer. Math. Soc.,* 115 (1965), 400-416.

[35] F. Riesz and B. Sz. Nagy, *Functional Analysis,* Ungar, New York, 1955.

[36] T. Ura, *Funkcial. Ekvac.* 2 (1959), 143-200.

[37] W. A. Coppel, *Stability and Asymptotic Behavior of Differential Equatio- s,* Heath, Boston, 1965.

INDEX